MW00910191

ANATOMY I LABORATORY MANUAL
2ND EDITION

BIOLOGY 233

GREGORY K. KARAPETIAN, PH.D.

HENRY FORD COLLEGE

Illustrations by Gregory Karapetian

ANATOMY I LABORATORY MANUAL. SECOND EDITION

Published by SK Scientific, LLC. PO Box 71274, Madison Heights, MI 48071. Copyright © 2013 by Gregory Karapetian. All rights reserved. No part of this publication may be reproduced or distributed in any form or by any means, or stored in a database or retrieval system, without the prior written consent of SK Scientific, LLC, including, but not limited to, in any network or other electronic storage or transmission, or broadcast for distance learning.

Library of Congress Cataloging-in-Publication Data
Karapetian, Gregory K.
Anatomy I Laboratory Manual / Gregory K. Karapetian – 2nd ed.
1. Human anatomy – Textbooks. 2. Human physiology – Textbooks.
Title: Anatomy I Laboratory Manual.

0 1 2 3 4 5 6 7 8 9

ISBN: 978-0-9889196-4-8

Proudly printed in the United States of America

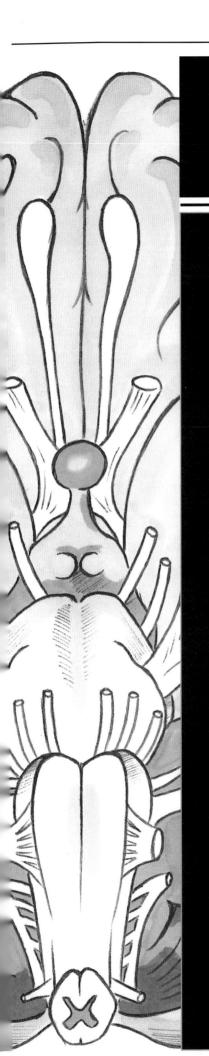

CONTENTS

BIOLOGY 233 AT A GLANCE

Biology 233 at Henry Ford College is the first semester class of the Anatomy & Physiology sequence, Biology 234 being the second. The laboratory component of the class is divided into three separate units – Unit I: bone anatomy; Unit II: muscle anatomy; and Unit III: neuroanatomy. Students enrolled in BIO 233 can anticipate a laboratory emphasis placed on proficiency in human anatomical structure.

Students are expected to be in lab every week, and should utilize the entire lab session to become proficient in identifying anatomical structures. Because labs are scheduled once per week, students are expected to commit to studying anatomy outside of the lab as well. The primary resource will be this lab guide. I have created this book to help students both in and out of the lab. The descriptions and illustrations will help facilitate the learning process, but simply purchasing the book will not guarantee success. Students need to understand that illustrations do not look exactly like models, and models (although wonderful teaching tools) do not look exactly like the real human body. Utilize the assigned textbook for the BIO 233 class – images may look different, and will thus offer another example of what you are learning.

Simply put, students must put time into this class. Anatomy and physiology is a challenging course, and lays the groundwork for those entering nearly any field in the Health Sciences. Having a sound and thorough understanding of the material will make for an easier time as you continue your education in a specialized field. Below is a short list of suggestions for doing well in the BIO 233 laboratory:

- Use the entire lab period. This is the time the models and your instructor are fully available to you – take advantage.

- Review, review, review. Everyone has their own learning style, and you must do what works for you; however repetition has long been one of the best ways for studying anatomy. In my opinion, there is no better substitution for studying anatomy than simply keeping the material in front of your eyes as much and as often as possible.

- Make a friend and work in groups. I know, I know – you already have enough friends on your online social network, but one or two more will not hurt… in fact, it will probably help. You will be with your classmates for the remainder of the semester so get used to challenging, teaching, and learning from each other – you will each benefit in the long run.

- In lab, do not immediately go to the instructor for help on every solitary item. The lab instructor is there to guide you, but not spoon feed the material. Learn to figure things out for yourself; this problem solving skill will be useful in the future. Of course the lab instructor can help, but do your best to look for the answer before you ask. There is no such thing as a dumb question, but be sure you have made an honest effort on your own to find the answer first.

- The best way to learn anatomy is through self discovery. Think of it this way: if you moved to a new city and did not know your way around, the best way to learn the area would be to get yourself lost – because once you are lost, you are forced to find your way back home. Learning anatomy is not much different. With each part of the body you study, you begin by looking at new roads, buildings, landmarks, etc… eventually you will put it all together. Will you make mistakes? Sure, but you would too in the "new city" scenario as well. It is okay to make mistakes, but learn from them – and better to make them sooner rather than later.

LABORATORY REGULATIONS

Proper attire is required. Although we do not perform "wet labs" or use chemicals in BIO 233, the college designates the room as laboratory space, and the State views it as such. For this reason long pants and shoes (covering toe to heel, including the top of the foot) must be worn at all times.

Handle all lab materials and models with care. **DO NOT** use a pencil or pen to point at the models, they are extremely expensive and pencil marks by one student at a time adds up over the course of a week, month, and semester. Treat the models with respect.

No food or drinks are allowed in the lab. No visitors are allowed in the laboratory.

Students may not remove any of the teaching materials from the lab. Models must be put back neatly after each lab. Clean up is not the instructor's main role, nor is it fair for one instructor to clean up after an entire class of students. A clean lab is the largely the responsibility of the student, please understand and respect this rule.

A NOTE FROM THE AUTHOR

Throughout this manual I have included sections called "Helpful Hints." My intention here is to step away from the science speak and communicate with you easy ways to remember concepts; I hope you find them useful.

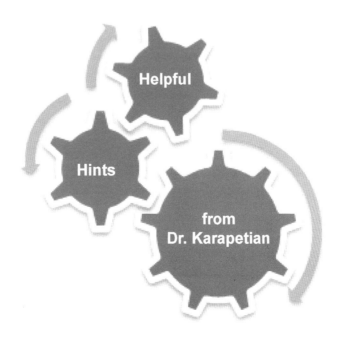

UNIT I

Introduction

While **physiology** explains the many different functions of the body, **anatomy** (*ana* = upward; *tome* = to cut) concentrates on the study of structure. Because the two disciplines are interrelated – we cannot completely separate structure from function – you will spend your laboratory time studying anatomy and class time on physiology. In order to help you better understand what you are learning I have included the roots of the words in parenthesis as you just saw above with the word 'anatomy.' Many words we use have a Greek or Latin base and learning them will be beneficial to you not only for BIO 233 and 234, but for those of you that plan to continue in the medical or allied health fields, you will see and use these terms on a daily basis. Many students find it a challenge to learn the new terms necessary for this class: sometimes spelling is not your strong suit, or perhaps English is not your first language. But like most of you, I do not speak fluent Latin or Greek – the languages in which many anatomical words are based. With study though, you can master these words too. Set your goals high, you can achieve them.

A brief history of anatomy

Dating back as far as 1600 BC, records indicate early Egyptians recognized specific structures within the body: the heart, liver, spleen, kidneys, bladder, and uterus. They even observed that blood vessels originate from the heart. The earliest anatomist we recognize today is **Hippocrates** (460 – 377 BC) whose studies demonstrated a basic comprehension of the musculoskeletal system, and the primitive descriptions of organ function. A large amount of his work, however, is believed to be based on hypotheses rather than actual observations of the body. **Aristotle** (384 – 322 BC) continued these efforts, on a much more scientific basis, by dissecting animals.

Human cadavers were first used for anatomical research when the Greek scientists **Herophilos** (335 – 280 BC) and **Erasistratus** (304 – 250 BC) were permitted to dissect the bodies of criminals in Alexandria. Herophilos, who you will remember as the developer of the Scientific Method, extensively recorded his findings and truly expanded the anatomical knowledge of the human body. These two men are also regarded as the founders of the legendary Medical School of Alexandria. The last notable anatomist of ancient times is **Galen** (131 – 201 AD), arguably the most accomplished of the ancient medical researchers.

Galen was a physician, surgeon, and philosopher who amassed the information by previous scholars and applied much of his own knowledge (by dissecting animals, as human specimens were unavailable) and created *the* anatomy textbook that was used for the next 1500 years.

Andreas Vesalius (1514 – 1564) is often credited with being the founder of modern human anatomy. Vesalius was an anatomist, physician, and the author of one of the most influential books on human anatomy, *De humani corporis fabrica* (*On the Fabric of the Human Body*). He refuted many past misconceptions about the human body by making direct observations and planned experiments.

Directional Terms

Anatomists utilize a set of **directional terms** (Table 1) that are precise, and allow us to avoid the use of unnecessary words.

Planes, Body Cavities

In human anatomy, studying the various regions and components can be complicated. In terms of spatial relations, if we did not have rules to follow, this task would be nearly impossible. When you give someone directions to your house, you use terms like "North" and "South." When you use a GPS navigation system, you hear words like "turn right" and "turn left." In the body we establish a frame of reference by using the **anatomical position**. Proper anatomical position is standing upright, chin and face forward, arms extended at the sides with open palms facing forward, and legs straight with toes pointing forward. From this position, we are able to view the entire front and back sides of an individual (Figure 1 & 2).

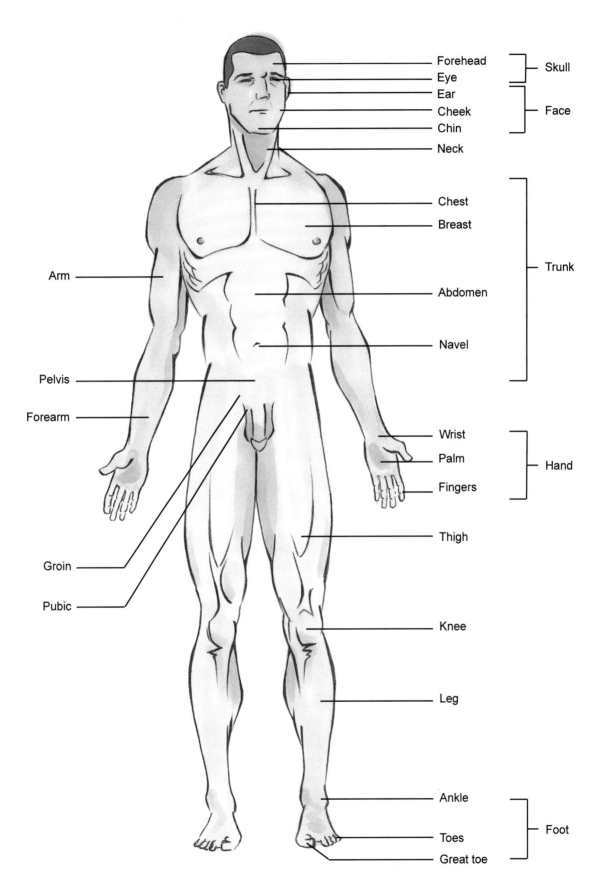

Figure 1. Anatomical position. Anterior view.

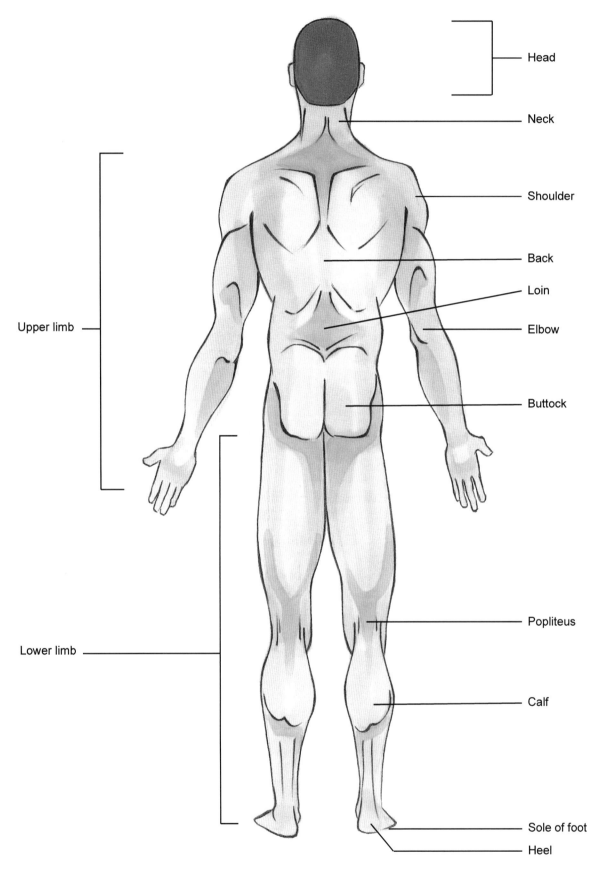

Head

Neck

Shoulder

Back

Loin

Elbow

Buttock

Upper limb

Popliteus

Lower limb

Calf

Sole of foot

Heel

Figure 2. Anatomical position. Posterior view.

Planes

Similar to the mathematical x, y, and z planes of geometry, **anatomical planes** are imaginary flat surfaces that pass through areas of the body. The **frontal** or **coronal** (*corona* = crown) **plane** is a vertical plane that separates the body into anterior and posterior sections. The **sagittal** (*sagittalis* = arrow) **plane** is also a vertical plane that divides the body into left and right sections. If this plane passes exactly through the midline of the body, thereby dividing it equally, it is referred to as the **midsagittal** (**median**) **plane**.

If the plane is off-center and divides the body into unequal left and right sides, it is referred to as a **parasagittal** (*para* = near) **plane**. If a plane divides the body into superior and inferior parts, it is called a **transverse** (**cross-sectional** or **horizontal**) **plane**. The frontal, sagittal, and transverse planes all exist at right angles to one another (Figure 3). There is a final plane, the **oblique plane**, that passes through the body at an angle.

Table 1. Directional Terminology

TERM	DEFINITION	EXAMPLE
Superior (cephalic or cranial)	Upper part of a structure; towards the head	The lungs are superior to the kidneys
Inferior	Lower part of a structure; away from head	The tibia is inferior to the femur
Caudal	Towards the tail	The hips are caudal to the waist
Anterior (ventral)	Front of the body	The heart is anterior to the spinal cord
Posterior (dorsal)	Back of the body	The spinal cord is posterior to the sternum
Medial	Near the midline of the body or a structure	The tibia is on the medial side of the leg
Lateral	Further from midline of body or a structure	The fibula is on the lateral side of the leg
Ipsilateral	On the same side of the midline of the body	The left lung and the spleen are ipsilateral
Contralateral	On opposite sides of the body's midline	The paired kidneys are contralateral
Proximal	Closer to the point of attachment	The elbow is proximal to the wrist
Distal	Further from the point of attachment	The wrist is distal to the elbow
Superficial	Toward the surface of the body	The skin is superficial to the internal organs
Intermediate	Between two structures	The heart is intermediate to the lungs
Deep	Away from the surface of the body	The liver is deep to the skin
Parietal	Forming the outer wall of a body cavity	The parietal pleura forms the outer layer of the pleural sacs that surround the lungs
Visceral	The covering of an organ within the ventral body cavity	The visceral pleura forms the inner layer of the pleural sacs and covers the external surface of the lungs

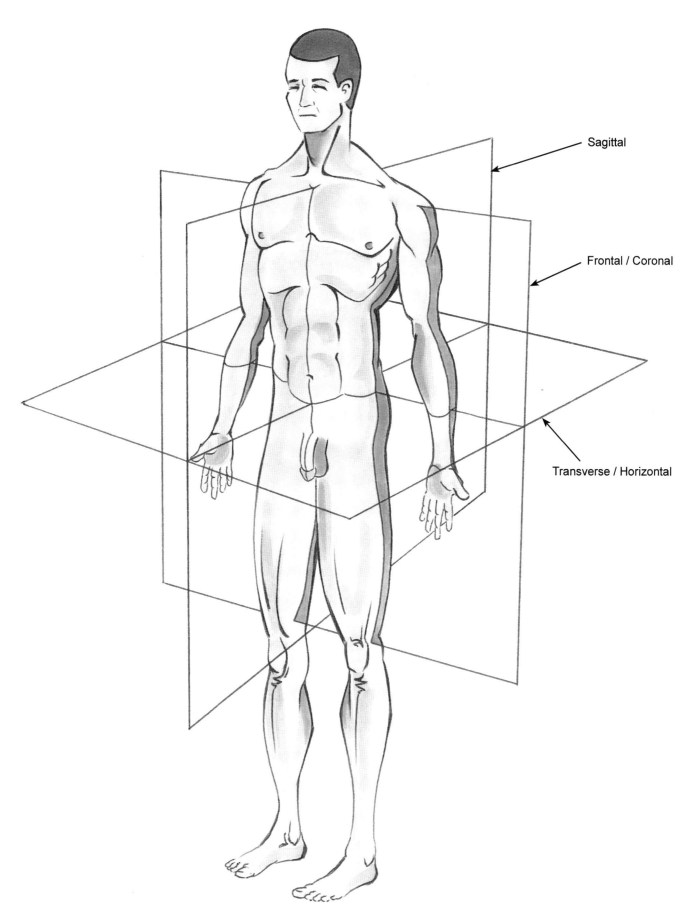

Sagittal

Frontal / Coronal

Transverse / Horizontal

Figure 3. Anatomical planes.

Body Cavities

Body cavities are spaces within each of us that serve to separate, support, and protect our internal organs. The two primary cavities are the dorsal and ventral body cavities (Figure 4). The **dorsal body cavity** lies to the posterior of the body. It can be further separated into the **cranial cavity** which contains the brain, and the **vertebral canal** which contains the spinal cord.

The second primary cavity is the **ventral body cavity**, located to the anterior of the body. Here, a thin **serous membrane** lines the walls and covers the organs (called **viscera** or **visceral organs**) inside. A serous membrane is a double-layered membrane that (1) lines a body cavity that does not open directly to the exterior and (2) covers the organs within that cavity. There are two main divisions of the ventral cavity as well: the superior **thoracic cavity** and the inferior **abdominopelvic cavity**. The **diaphragm** (*diaphragma* = dividing wall), an important muscle used for breathing, is the structure that divides the ventral cavity into upper and lower parts.

The thoracic cavity is comprised of three sections: two pleural cavities and one pericardial cavity. Each **pleural** (*pleur* = rib, side) **cavity** is a small, fluid filled space between the part of the serous membrane that covers the lung and the part that lines the wall of the thoracic cavity. The **pericardial cavity** is the space that surrounds the heart.

As the name implies, the **abdominopelvic cavity** is divided into two parts: the superior portion is the **abdominal cavity** (*abdere* = to hide, as in hiding the viscera) which contains the stomach, liver, gallbladder, pancreas, spleen, and the majority of the small intestine. Lining the abdominal cavity is a serous membrane that covers the organs which is called the **peritoneum**. The inferior portion of the abdominopelvic cavity is the **pelvic cavity**, which contains the lower sections of the large intestine, the urinary bladder, and the (internal) reproductive organs.

Abdominal Regions

Anatomists commonly use a nine region division when describing the organs found in the abdominopelvic cavity (Figure 5). This is a detailed system of organization that allows anatomists to be specific in their description of organ location. Collectively these nine regions are called the **abdominopelvic regions**; they can be seen in Table 2. More simply, the abdominopelvic cavity can divided into **quadrants** (*quad* = four). Imagine a horizontal and vertical line crossing and making a " + " at the navel. These intersecting lines divide the abdomen into a **right upper quadrant** (**RUQ**), **left upper quadrant** (**LUQ**), **right lower quadrant** (**RLQ**), and **left lower quadrant** (**LLQ**). Those of you doing clinical work will more often use the quadrant system when describing the site of an abdominopelvic abnormality or pain – it is a quick and simple system to use.

Table 2. Abdominopelvic Regions

REGION	DEFINITION
Right hypochondriac	*hypo* = under; *chondro* = cartilage
Epigastric	*epi* = above; *gaster* = stomach
Left hypochondriac	
Right lumbar	*lumbus* = loin
Umbilical	*umbilikus* = navel
Left lumbar	
Right iliac (inguinal)	iliac refers to superior part of hipbone
Hypogastric (pubic)	*hypo* = under; *gaster* = stomach
Left iliac (inguinal)	

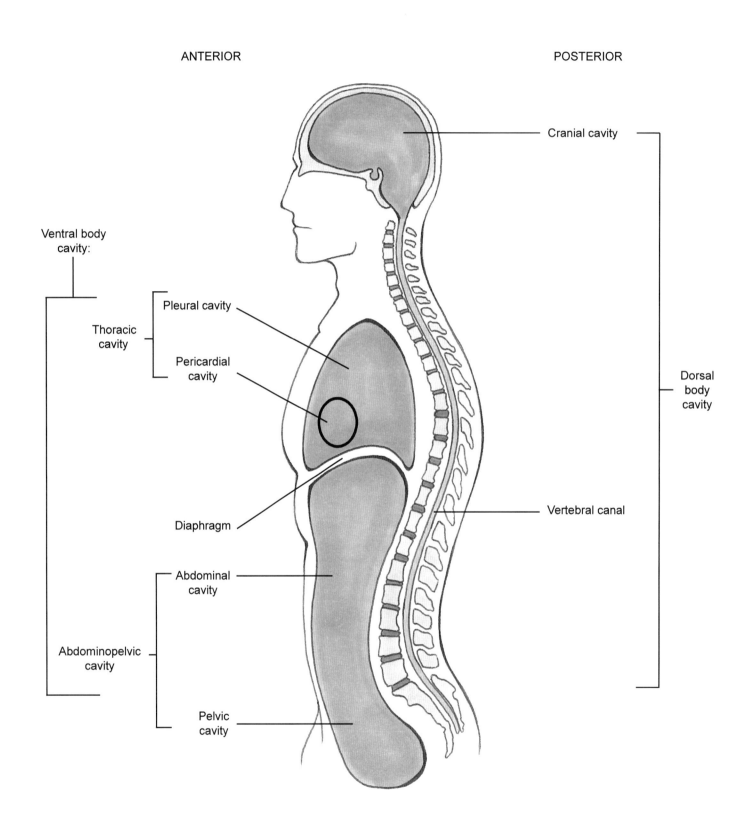

ANTERIOR

POSTERIOR

Cranial cavity

Ventral body
cavity:

Pleural cavity

Thoracic
cavity

Pericardial
cavity

Dorsal
body
cavity

Diaphragm

Vertebral canal

Abdominal
cavity

Abdominopelvic
cavity

Pelvic
cavity

Figure 4. Body cavities. Midsagittal view.

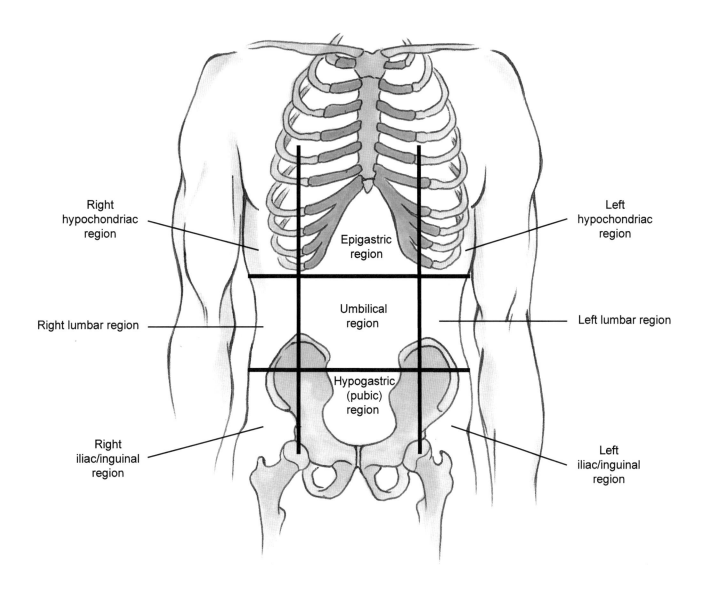

Figure 5. Abdominopelvic cavity. Anterior view.

Bone Introduction

he average human body has 206 bones (Figure 6 & 7); but keep in mind each of us probably has a number of small sutural bones that may be present in the skull, so the number 206 can exist within a small range. When we study the bones we categorize them as either part of the axial skeleton or the appendicular skeleton.

The **axial skeleton** consists of the bones that form the long axis of the body and provide support and protection for organs of the head, neck, and trunk (Table 3).

1. **Skull** – consists of two sets of bones: the cranial bones that form the cranium and the facial bones which support the eyes, nose, and jaw.

2. **Auditory ossicles** – three bones (that transmit sound impulses) are present on each side of the skull within the middle-ear chamber.

3. **Hyoid bone** – located inferior to the jaw and superior to the larynx, this bone supports the tongue and assists in swallowing.

4. **Vertebral column** – consists of 26 individual bones separated by (cartilaginous) intervertebral discs. In the lower portions, several bones fuse together to form the sacrum and a few more fuse to form the coccyx (tailbone).

5. **Thoracic cage** – commonly referred to as the rib cage, forms the framework of the thorax. The rib cage articulates with the thoracic vertebrae on the posterior side of the body and includes 12 pairs of ribs. The costal cartilage connect the ribs to the sternum on the anterior side of the body.

The **appendicular skeleton** is comprised of bones of the upper and lower extremities and bony girdles that anchor the limbs to the axial skeleton (Table 4).

1. **Pectoral girdle** – consists of the paired scapulae and clavicles. This girdle serves as the attachment point for the muscles that move the upper arm and forearm. Interestingly, the only direct bony connection point for the entire arm occurs at this area when the clavicles articulate with the sternum.

2. **Upper extremities** – the brachium (upper arm) is made up of a solitary humerus, while the antebrachium (forearm) consists of a radius and ulna. The wrist is made up of carpal bones, while the hand is comprised of metacarpal bones and phalanges. These bones are each mirrored on the opposite arm.

3. **Pelvic girdle** – formed by two os coxae (hip bones), and joined anteriorly by the pubic symphysis and posteriorly by the sacrum. The pelvic girdle offers protection of the lower visceral organs and supports the weight of our bodies.

4. **Lower extremities** – the femur is the bone of the thigh; the tibia and fibula are the bones of the leg; and the tarsal bones, metatarsal bones, and the phalanges all comprise the foot. The patella is the "knee cap" located between the thigh and leg on the anterior side. These bones are each mirrored on the opposite leg.

Classification of the Bones of the Adult Skeleton

Table 3. Divisions of the Adult Skeletal System: **AXIAL**

REGIONS OF THE SKELETON	NUMBER OF BONES
AXIAL SKELETON	
Skull	
Cranial bones	8
Facial bones	14
Auditory ossicles	6
Hyoid bone	1
Vertebral column	26
Thorax	
Sternum	1
Ribs	24
SUBTOTAL	**= 80**

Table 4. Divisions of the Adult Skeletal System: **APPENDICULAR**

REGIONS OF THE SKELETON	NUMBER OF BONES
APPENDICULAR SKELETON	
Pectoral (shoulder) girdles	
Clavicle	2
Scapula	2
Upper limbs (extremities)	
Humerus	2
Radius	2
Ulna	2
Carpals	16
Metacarpals	10
Phalanges	28
Pelvic (hip) girdle	
Os coxae (hip bones)	2
Lower limbs (extremities)	
Femur	2
Tibia	2
Fibula	2
Patella	2
Tarsals	14
Metatarsals	10
Phalanges	28
SUBTOTAL	**= 126**
TOTAL 80 (Axial) + 126 (Appendicular)	**= 206**

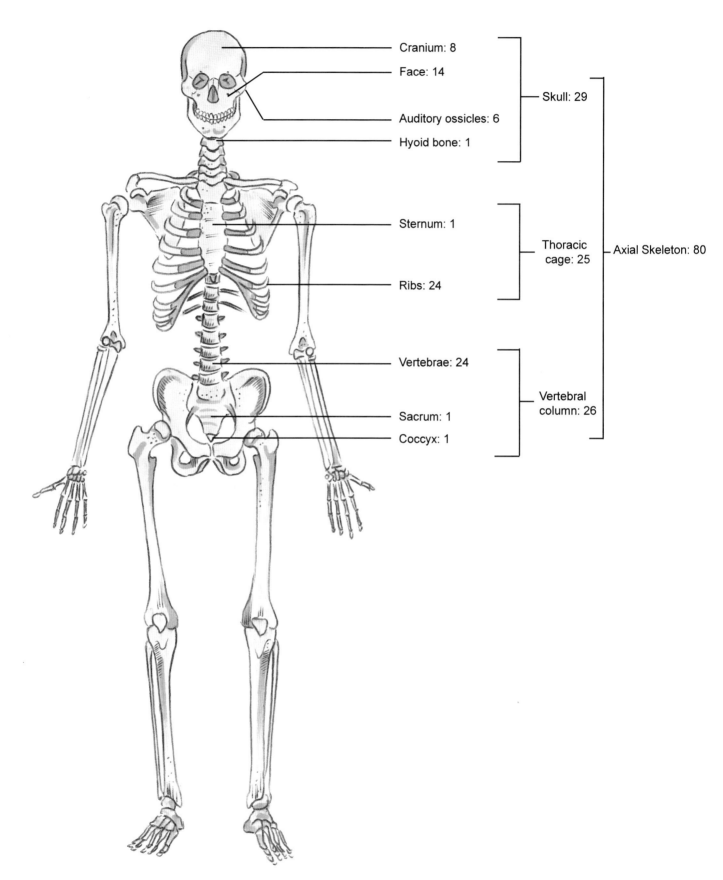

Figure 6. Human skeleton. Anterior view, axis bones are indicated.

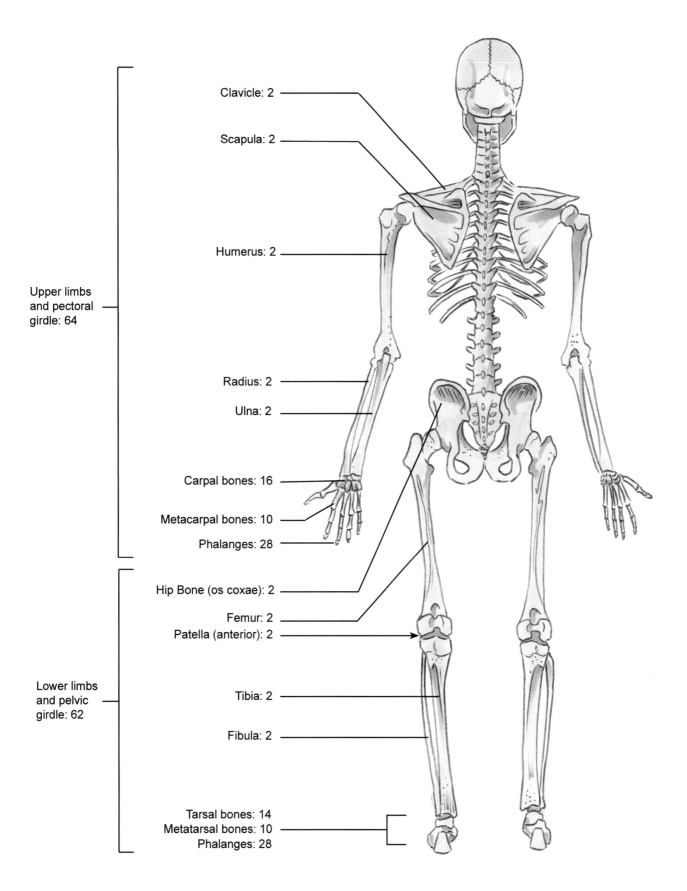

Clavicle: 2

Scapula: 2

Humerus: 2

Upper limbs and pectoral girdle: 64

Radius: 2

Ulna: 2

Carpal bones: 16

Metacarpal bones: 10

Phalanges: 28

Hip Bone (os coxae): 2

Femur: 2

Patella (anterior): 2

Lower limbs and pelvic girdle: 62

Tibia: 2

Fibula: 2

Tarsal bones: 14
Metatarsal bones: 10
Phalanges: 28

Figure 7. Human skeleton. Posterior view, appendicular bones are indicated.

Before you commit to learning the bones, it is worth spending time learning the terminology in Table 5 beforehand. Having a solid understanding of these terms will prove to be extremely useful when the time comes to recall all the detailed, component parts of each bone.

Axial Skeleton

Remember when you were a kid spinning a globe as fast as you could and stopping it with your finger as you announced, "I'm going to go on vacation... here!" then pointing to some random country? Well, the earth rotating on that imaginary line from the North to South Pole was its axis. In the same way, we too have an imaginary line running the length of our torso. We call this our axial skeleton; and by definition, an **axis** is a straight line through a body or the center around which something rotates. When we consider our own axis from a skeletal standpoint, the structures included are the ones that fit with this definition.

We will start the study of the axial skeleton with the skull, followed by the vertebral column, and then the thoracic cage – all of which are, anatomically, straight up and down the middle of our bodies.

Adult Skull

The adult skull consists of 22 bones that can be divided into two separate sets: cranial bones and facial bones (Table 6). The **cranial** (*cranium* = brain case) **bones** offer protection to the brain by forming the cranial cavity. The eight cranial bones are the frontal bone (1), parietal bones (2), temporal bones (2), sphenoid bone (1), ethmoid bone (1), and the occipital bone (1). There are 14 **facial bones** that combine to form the face; the nasal bones (2), maxillae (2), lacrimal bones (2), zygomatic bones (2), inferior nasal conchae (2), vomer (1), palatine bones (2), and mandible (1). In addition to forming the large cranial cavity, the bones of the skull also form smaller cavities – the nasal cavity and the orbits (eye sockets). Smaller still are the paranasal sinuses, which are mucous lined cavities that open into the nasal cavity. Additionally, there are tiny cavities within the skull that house the component parts necessary for hearing and balance.

Table 5. Surface features of bones

STRUCTURE	DESCRIPTION	EXAMPLE
Articulating Surfaces		
Condyle	Large, rounded, articulating knob	Occipital condyle of occipital bone
Facet	Flattened or shallow articulating surface	Costal facet of a thoracic vertebra
Head	Prominent, rounded, articulating end of a bone	Head of the femur
Depressions and Openings		
Alveolus	Deep pit or socket	Alveoli for teeth in the mandible
Fissure	Narrow, slitlike opening	Superior orbital fissure of the sphenoid bone
Foramen	Rounded opening through a bone	Foramen magnum of the occipital bone
Fossa	Flattened or shallow depression on bone surface	Mandibular fossa of the temporal bone
Sinus	Cavity or hollow space in a bone	Frontal sinus of the frontal bone
Sulcus	Groove that contains a tendon, nerve, or vessel	Intertubercular sulcus of the humerus
Nonarticulating Structures		
Crest	Narrow, ridgelike projection	Iliac crest of the os coxae
Epicondyle	Projection superior to a condyle	Medial epicondyle of the femur
Process	Projection or outgrowth	Zygomatic process of the temporal bone
Spine	Sharp, slender process	Spine of the scapula
Trochanter	Massive process found only on the femur	Greater trochanter of the femur
Tubercle	Small, rounded process	Greater tubercle of the humerus
Tuberosity	Large, roughened process	Radial tuberosity of the radius

Table 6. Bones of the Adult Skull

CRANIAL BONES	NUMBER	FACIAL BONES	NUMBER
Frontal	1	Nasal	2
Parietal	2	Maxillae	2
Temporal	2	Lacrimal	2
Sphenoid	1	Zygomatic	2
Ethmoid	1	Inferior nasal conchae	2
Occipital	1	Vomer	1
		Palatine	2
		Mandible	1
SUBTOTAL	= 8	SUBTOTAL	= 14
		TOTAL	= 22

ost of the bones of the skull do not move (the exceptions being the mandible and the auditory ossicles) and are held together by an immovable joint called a **suture** (SUE-cher; *sutura* = seam). Sutures are only found between the bones of the skull and interlock like a jigsaw puzzle – a design that offers strength and reduces the chance of fracturing.

Four of the prominent sutures of the adult skull include the:

1. **Coronal** (*corona* = crown) **suture** – unites the frontal bone and the two parietal bones.

2. **Sagittal** (*sagitta* = arrow) **suture** – unites the two parietal bones. Named so because this suture and its associated fontanels (soft spots), in infants, resemble an arrow.

3. **Lambdoid suture** – unites the parietal and occipital bones. Named after the Greek letter lambda (Λ) in which it resembles.

4. **Squamous** (*squama* = flat) **suture** – unites the parietal and temporal bones. Named because the part of the temporal bone that joins with the parietal bone to form this suture is a thin, flat region called the **temporal squama** or **squamous area of temporal bone**.

The skull consists of the cranium and the mandible. The posterior aspect of the cranium is termed the **neurocranium**, and as the name implies, it is the part that protects the brain. The anterior/inferior aspect of the cranium is the **facial cranium** where the facial bones form the contours of our faces.

Cranial Bones

There are eight cranial bones. Their detailed descriptions are as follows:

Frontal Bone (1)

The forehead, the roofs of the orbits (eye sockets), and the majority of the anterior portion of the cranial floor is formed by the **frontal bone**, the most anterior bone of the upper cranium (Figure 8 & 9). Superior to the orbits, the frontal bone begins to thicken and form the **supraorbital** (*supra* = above; *orbital* = orbit) **margin**. Within the supraorbital margin is a hole called the **supraorbital foramen** in which nerve and blood supply to the eyebrows and forehead pass. A slight variation on the supraorbital foramen is the **supraorbital notch** which can be seen in some people and serves the same purpose. The **frontal sinuses** lie just deep to the frontal squama – a flattened area which creates the forehead.

Parietal Bones (2)

The two **parietal** (*paries* = wall) **bones** form the largest parts of the sides and roof of the cranium, and contain many projections and depressions that house the blood vessels that supply the outer coverings of the brain (Figure 8).

= bones   = foremen/ holes   = suture 4

Right   = random   Left.

Sutures:
C S S L
o q q a
r u u m
o a a b
n m m d
a o o o
l u u i
    s s d

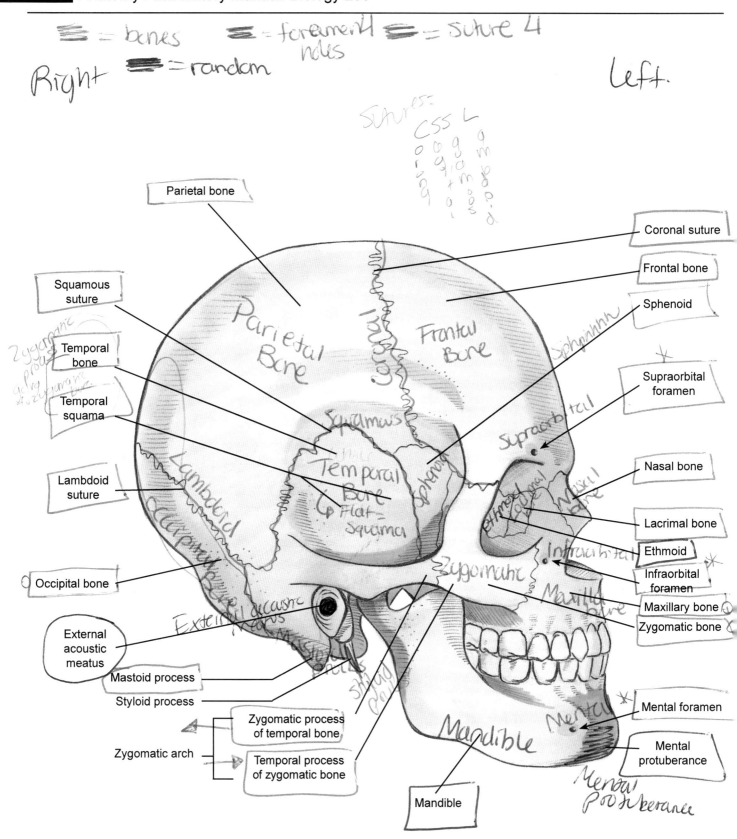

Parietal bone

Coronal suture

Frontal bone

Sphenoid

Supraorbital foramen

Squamous suture

Temporal bone

Temporal squama

Nasal bone

Lacrimal bone

Ethmoid

Infraorbital foramen

Lambdoid suture

Maxillary bone

Zygomatic bone

Occipital bone

External acoustic meatus

Mastoid process

Styloid process

Zygomatic process of temporal bone

Temporal process of zygomatic bone

Zygomatic arch

Mandible

Mental foramen

Mental protuberance

Figure 8. Adult skull. Lateral view.

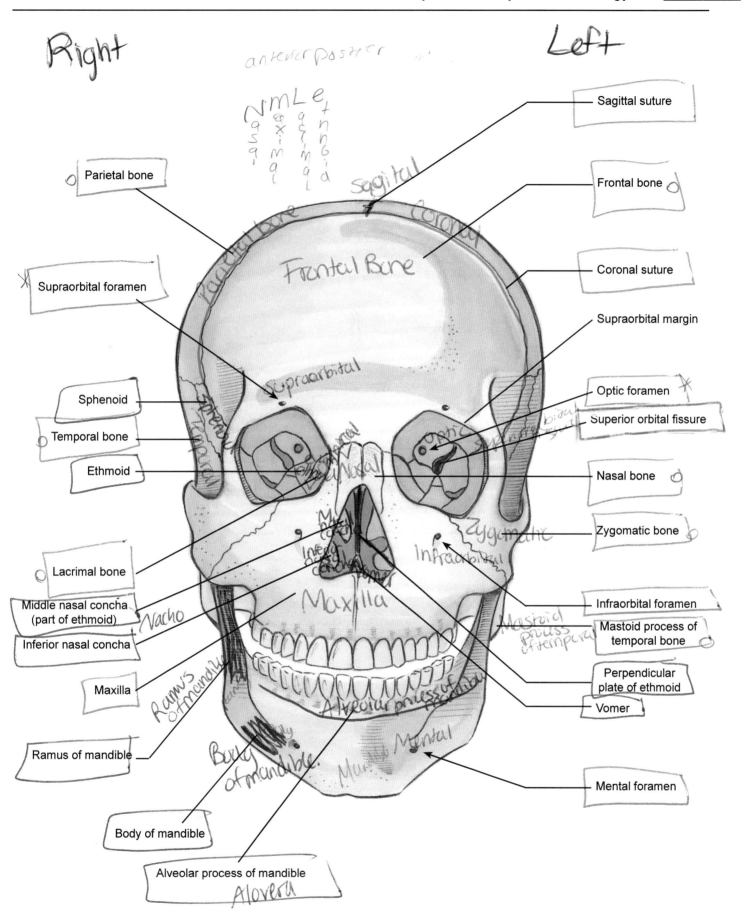

Figure 9. Adult skull. Anterior view.

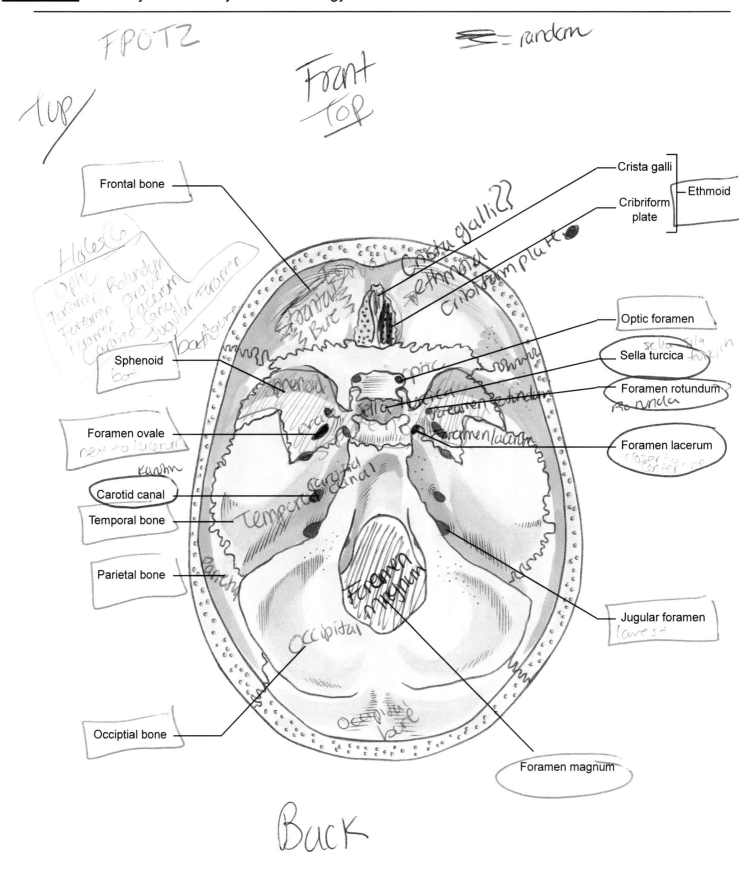

Figure 10. Adult skull. Horizontal section.

Temporal Bones (2)

The two **temporal** (*tempora* = temples) **bones** form the inferior and lateral aspects of the cranium (Figure 8). The **zygomatic process** projects from the squamous area of the temporal bone (temporal squama) and contributes to the zygomatic arch that forms part of our cheek bones. On the inferior surface of the zygomatic process of the temporal bone are two notable structures: the **mandibular (glenoid) fossa** and the articular tubercule, both of which articulate with the condylar process of the mandible (the lower jaw) to form the temporomandibular joint (TMJ).

The lower internal structures of the temporal bone accommodate space for the internal and middle ear, and contain the **carotid canal** and **jugular foramen** (Figure 10). The **external auditory (acoustic) meatus**, the **mastoid** (*mastoid* = breast shaped) **process**, and **styloid** (*styloid* = stake) **process** can all be observed from the lateral view of the skull (Figure 8). These structures form the ear canal and serve as points of attachment of neck muscles and muscles of the tongue, respectively.

Sphenoid Bone (1)

The **sphenoid** (SFĔ-noyd; *spheno* = wedge shaped) **bone**, as the name implies, is wedged between several bones in the middle of the skull. When you look at the floor of the cranium from an above view, note that the sphenoid bone articulates with the frontal bone (anteriorly), the temporal bones (laterally), and the occipital bone (posteriorly) (Figure 10 & 8). The sphenoid bone is sometimes described as "bat-shaped" with the **greater wings** forming the anteriolateral floor of the cranium. The **lesser wings** are smaller and form the ridge located anterior and superior to the greater wings. In the middle of the bone, on the superior side, is a depression called the **sella turcica** (*sella turcica* = Turkish saddle), which houses and protects the pituitary gland. The **optic** (*optikus* = eye) **foramen** and the **superior orbital fissure** lie within the sphenoid bone.

Ethmoid Bone (1)

The **ethmoid** (*ethmos* = sieve) **bone** lies on the cranial floor, between the eye sockets, and supports the nasal cavity (Figure 10). It is anterior to the sphenoid bone and posterior to the nasal bones.

Because it sits in the midline, the **perpendicular plate** of the ethmoid bone forms the superior part of the **nasal septum** (*saeptum* = wall, partition) while the **cribriform plate** forms the roof of the nasal cavity, and contains many **olfactory** (*olfacere* = to smell) **foramina**. A triangular process, the **crista galli** (*crista galli* = cock's comb), juts out superiorly from the cribriform plate and acts as a point of attachment for the meninges (membranes) that cover the brain. There are also many open cavities, or sinuses (hence, *ethmos*) within this bone.

The ethmoid bone forms projections on either side of the nasal septum called the **superior nasal concha** (KONG-ka; plural KONG-kē; *concha* = shell) and the **middle nasal concha**. These structures create turbulence and cause air to swirl in the nasal cavity when we inhale so they are sometimes called **turbinate bones**. Breathing through the nose helps to trap particles and debris on this mucus lined area.

Occipital Bone (1)

The **occipital** (*occipital* = back of head) **bone** forms the most posterior portion of the cranium (Figure 8 & 11). It contains a large hole, the **foramen magnum**, which is the opening in which the brain connects to the spinal cord. It is here where the medulla oblongata and the vertebral arteries also pass through. The **external occipital protuberance (inion)** is a noticeable projection located superior and posterior to the foramen magnum. You may be able to palpate this on the lower back side of your own head. On either side of the foramen magnum are smooth, oval shaped processes called the **occipital condyles**, the articulation point of the skull with the first cervical vertebra. This union is known as the atlanto-occipital joint.

Facial Bones

ourteen bones make up the facial cranium: The (2) nasal bones form the upper portion of the bridge of the nose; the (2) maxilla bones form the upper jaw and contribute to the orbits; (2) lacrimal bones are located in each orbit (eye socket) next to the nose and near the tear ducts; the (2) zygomatic bones form the lateral and inferior portion of the orbits and help to form the cheek structure; there are (2) inferior nasal conchae (inferior turbinates) of the nose; the (1) vomer, along with a portion of the ethmoid bone, forms the nasal septum; the hard palate is comprised of (2) palatine bones; and finally, the (1) mandible forms the lower jaw.

Nasal Bones (2)

The bridge of the nose is formed by the two **nasal bones** (Figure 9). The remainder of the nose is formed by cartilage.

Maxillae (2)

There are two **maxilla bones** that form the upper jawbone, areas of the nasal cavity, the inferior parts of the orbits, and the majority of the hard palate (Figure 8, 9 & 11). Within the maxillae are the **alveolar** (*alveolus* = hollow) **processes**, arches that contain the sockets for the upper teeth. Inferior to each eye socket are the **infraorbital** (*infra* = below; *orbital* = orbit) **foramen**. The maxillae articulate with every bone of the face except the mandible.

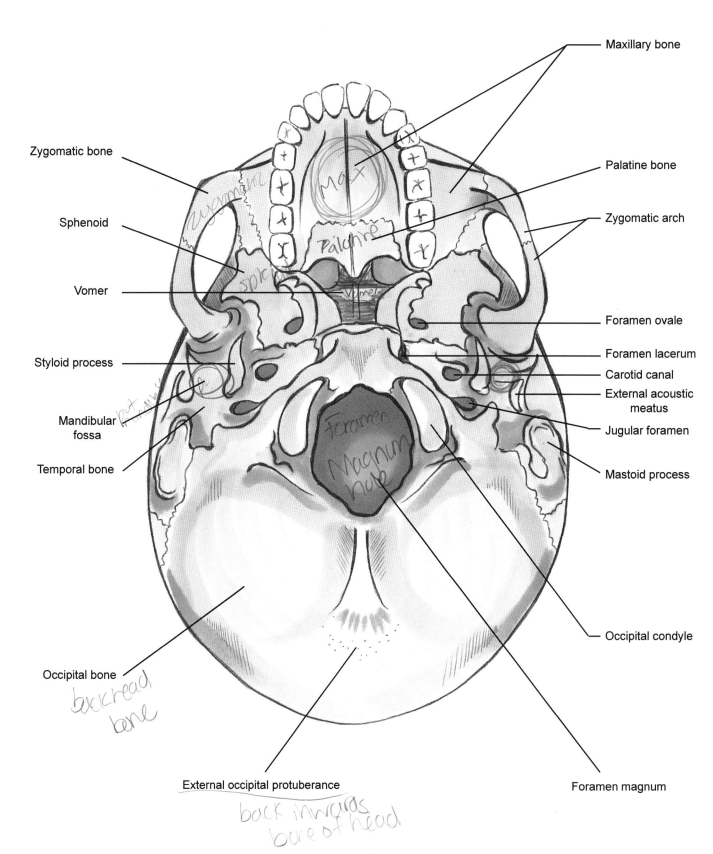

Figure 11. Adult skull. Inferior view.

Lacrimal Bones (2)
The smallest bones of the face are the **lacrimal** (*lacrima* = tear) **bones**, found posterior and lateral to the nasal bones (Figure 8). The lacrimal bones are situated in the medial walls of each orbit, and have structures that help collect tears and send them to the nasal cavity.

Zygomatic Bones (2)
The bones we commonly recognize as the "cheekbones" actually extend upward from the cheek to form the lateral wall, and inward to form the floor, of the eye sockets – they are the **zygomatic** (*zygoma* = cheekbone) **bones** (Figure 8 & 9). The **zygomatic arch** is formed by the union of the zygomatic process of the temporal bone and the temporal process of the zygomatic bone.

Inferior Nasal Conchae (2)
There are two **inferior nasal conchae** (KONG-kē) or **turbinates** that extend medially into the nasal cavity (Figure 9). Like the superior and middle nasal conchae, they promote circulation of air within the nasal cavity. Note: these bones are not part of the ethmoid bone, as the superior and middle nasal conchae are.

Vomer (1)
Forming the posterior and inferior part of the nasal septum is the **vomer** (*vomer* = plowshare) (Figure 9).

Palatine Bones (2)
The **palatine** (*palatum* = plate) **bones** form the posterior segment of the hard palate and part of the nasal cavity (Figure 11). The horizontal plates of the palatine bones, because they are L-shaped, separate the nasal from the oral cavity.

Mandible (1)
The **mandible** (*mandere* = to chew) is the largest and most powerful of the facial bones and is, other than the auditory ossicles, the only real movable bone of the skull (Figure 8, 9 & 12). Extending from the base of the skull the mandible has, on each side, a **condylar process** that articulates with the **mandibular fossa** of the temporal bone to form the **temporomandibular joint (TMJ)**.

On the superior/anterior side there is also a **coronoid process**. Each **ramus** (*ramus* = branch) then extends downward, reaching an **angle**, and lengthens forward to form the **body**. Similar to the maxillae, the **alveolar** (*alveolus* = hollow) **processes** are arches that contain the sockets for the lower teeth. The **mental** (*mental* = chin) **foramen** is the opening for the mental nerve.

Before moving on to study the vertebral column, there are a few key structures of the upper axial skeleton worth noting. They are foramina, fontanels, and the hyoid bone.

Foramina
Foramina are the passageways for blood and nerve supply to the brain and various structures of the skull.

Fontanels
Embryonic skeletons are largely made up of cartilage and fibrous connective tissues. Before the replacement of these structures occurs (ossification), **fontanels** (*fontanels* = little fountains) are present between cranial bones. The four main fontanels are the: **anterior fontanel**, **anteriolateral (sphenoidal) fontanel**, **posterior (occipital) fontanel**, and the **posteriolateral (mastoid) fontanel** (Figure 13). These structures are commonly called "soft spots" and allow for easier passage through the birth canal, as well as rapid growth of the brain during early childhood.

Hyoid Bone
The **hyoid** (*hyoedes* = U-shaped) **bone** is not part of the skull or vertebral column, but because of its location, is often studied at the same time (Figure 14). Interestingly, the hyoid bone does not articulate with any other bone – it is held in place by muscles and ligaments. It is located inferior to the mandible, and superior to the larynx. The hyoid bone provides an attachment point for muscles of the neck and helps to support the tongue.

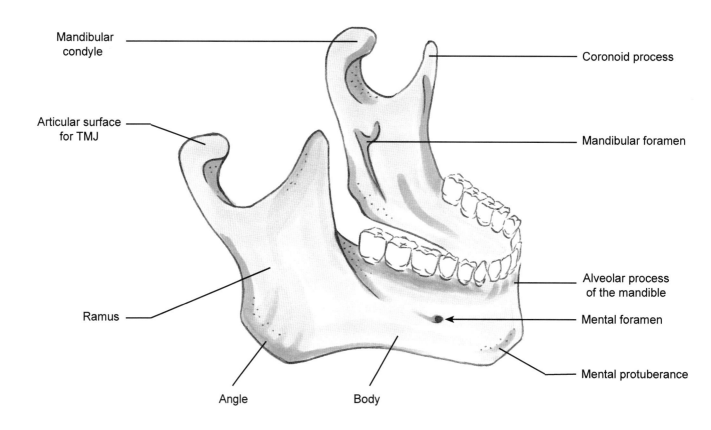

Figure 12. Mandible.

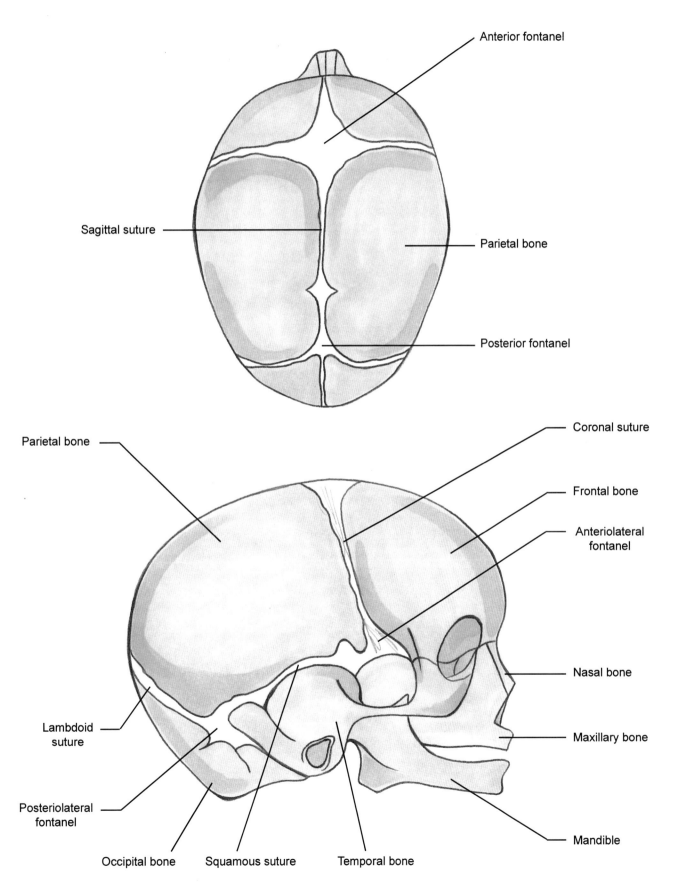

Figure 13. Skull of an infant. Superior and lateral view.

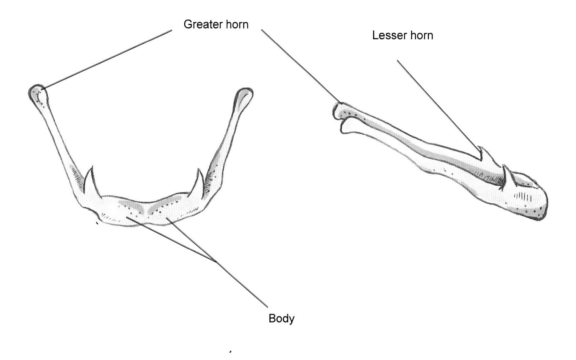

Figure 14. Hyoid bone. Anterior and lateral view.

Vertebral Column

As we consider the long axis of our bodies, the axial skeleton continues down from the skull as the vertebral column (Figure 15). Essentially a chain of bones, the vertebral column is divided into five regions: there are seven cervical vertebrae (C1 – C7); twelve thoracic vertebrae (T1 – T12) all of which have ribs attached to their sides; five lumbar vertebrae (L1 – L5); the one sacrum is formed by five fused bones; and one coccyx is formed by the fusion of (typically) four bones.

The vertebral column achieves its overall flexibility by an additive effect. Consider what happens when we bend over to touch our toes. This is not like the hinge of an oven door opening and closing, rather a sum of many small movements that occur between each solitary vertebra. When all of these individual actions take place, it gives the appearance of one large movement. The vertebrae enclose and protect the spinal cord, support the skull, and provide an attachment point for the rib cage and torso muscles. When the vertebrae are stacked on top of one another, openings called the **intervertebral foramina** form, which allow for the passage of spinal nerves (Figure 16).

Most vertebrae are generally similar in their structure; they usually contain an anterior oval or circular shaped body that comes in contact with the **intervertebral discs** on their superior and inferior sides. The **vertebral arch** projects from the posterior side and is made up of two supporting **pedicles** (*pediculus* = small foot) and two arched **laminae** (*lamina* = thin layer). This semicircular structure creates an opening called the **vertebral foramen**, through which the spinal cord passes.

Seven processes arise from the vertebral arch: two **transverse processes** extend laterally from each side of a vertebra; a **spinous process** protrudes posteriorly and inferiorly from the vertebral arch; and two **superior articular processes** have interlocking articulations with the two **inferior articular processes**.

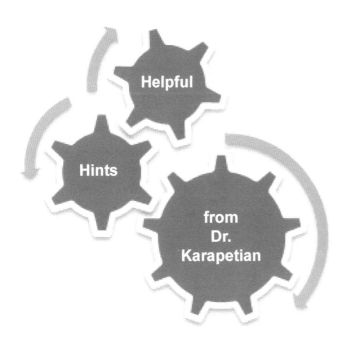

To remember the numbers of bones in each of the spinal regions think of the times of the day we eat our meals: Breakfast at 7:00, Lunch at 12:00, Dinner at 5:00. That's 7 cervical vertebrae, 12 thoracic vertebrae, and 5 lumbar vertebrae – easy!

7 cervical vertebrae
12 thoracic vertebrae
5 lumbar vertebrae

Cervical Vertebrae

Atlas (C1) and Axis (C2)

The atlas and axis (*axis* = axle) are unique vertebrae (Figure 17). The **atlas** (**C1**) does not have a body or spinous process like the other vertebrae; it is essentially a ring-like bone. When we nod our heads "yes," there is a sliding movement as the atlas forms the **atlanto-occipital joint** with the base of the skull – remember the occipital condyles? The **axis** (**C2**) forms a pivoting **atlantoaxial** joint with the atlas. This pivoting movement can occur because of a bony projection called the **dens**. It is this articulation formed at the midline that moves when we shake our heads "no" from side to side.

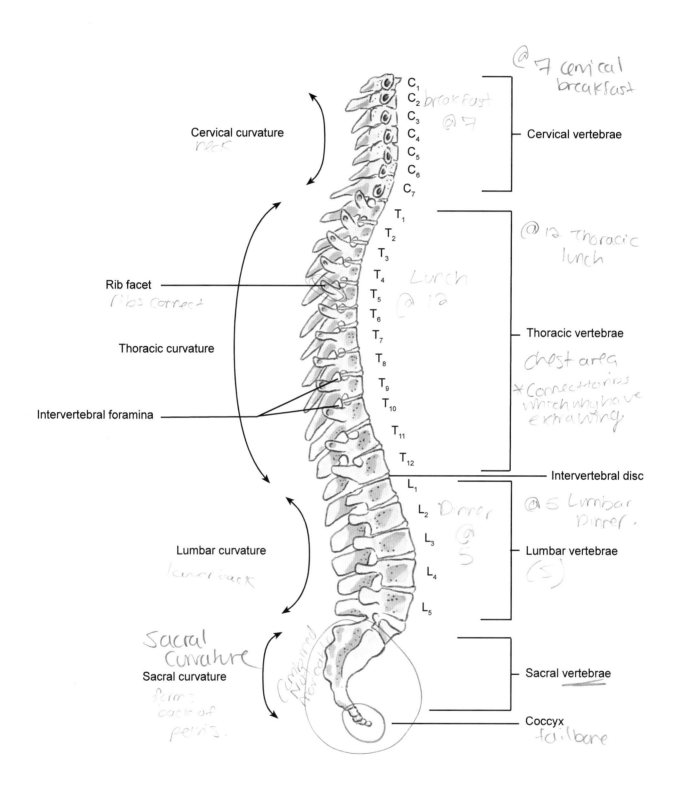

Cervical curvature
neck

Rib facet
ribs connect

Thoracic curvature

Intervertebral foramina

Lumbar curvature
lower back

Sacral
curvature

Sacral curvature
forms
back of
pelvis.

C₁
C₂ breakfast
@ 7
C₃
C₄
C₅
C₆
C₇

T₁
T₂
T₃
T₄
T₅
T₆
T₇
T₈
T₉
T₁₀
T₁₁
T₁₂

Lunch
@ 12

L₁
L₂ Dinner
@
5
L₃
L₄
L₅

Combined not called

@ 7 cervical
breakfast

Cervical vertebrae

@ 12 Thoracic
lunch

Thoracic vertebrae

chest area
* Connect to ribs
which why have
extra wing

Intervertebral disc

@ 5 Lumbar
Dinner.

Lumbar vertebrae
(5)

Sacral vertebrae

Coccyx
tailbone

Figure 15. Vertebral column. Lateral view.

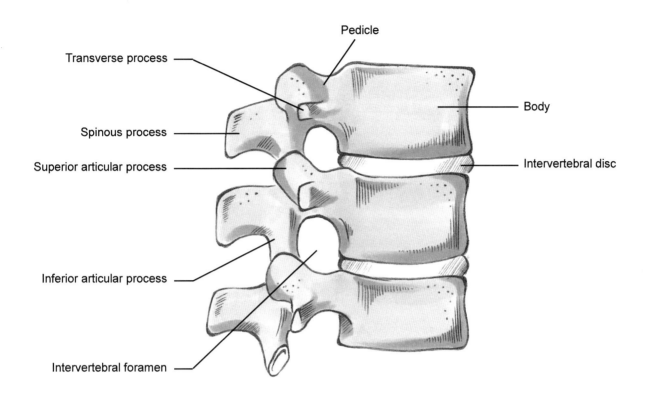

Figure 16. Vertebral anatomy. Lateral view.

he remaining **cervical vertebrae** (C3 – C7) are similar to each other in the following structures: they each have a **body**, a **vertebral arch** made up of the **pedicles** and **laminae**, and a component unique to cervical vertebrae – the paired **transverse foramen**. The cervical vertebrae have a pair of **superior articular processes** (with **facets**; *facette* = little face) and **inferior articular processes** (with facets) which project up and down from the right and left sides of the vertebral arch (Figure 18). These form facet joints with the articular processes of the bones immediately above and below them.

Thoracic Vertebrae

Thoracic vertebrae are larger than cervical vertebrae and continue to increase in size from T1 to T12 (Figure 19). Their spinous processes slope downward and they have paired **costal facets** that allow for articulation with the ribs. We do not see rib articulation in the other regions of the vertebral column.

Lumbar Vertebrae

Lumbar (*lumbus* = loin) **vertebrae** are the largest of the vertebrae and can be identified by their weighty bodies and thick spinous processes, ideal attachment points for the large muscles of the back. The lumbar vertebrae have distinct superior articular processes that face medially instead of superiorly, and inferior articular processes that face laterally instead of inferiorly (Figure 20).

The **vertebral foramina** change shape as the spinal cord continues down the vertebral column. In the cervical region each vertebral foramen tends to be large and triangular, in the thoracic region they are more circular, and in the lumbar region they are relatively small and triangular (Figure 21).

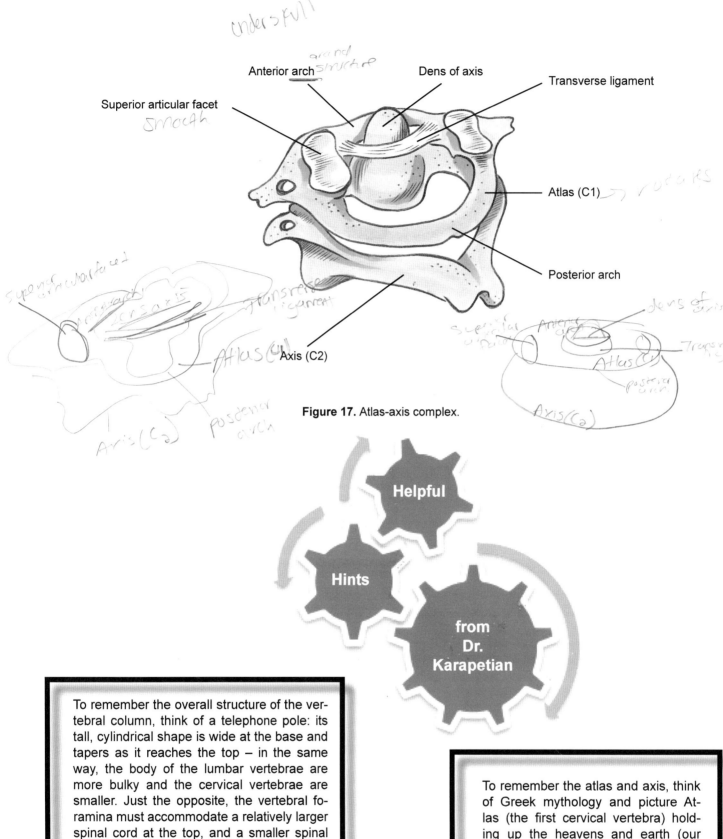

Figure 17. Atlas-axis complex.

Helpful Hints from Dr. Karapetian

To remember the overall structure of the vertebral column, think of a telephone pole: its tall, cylindrical shape is wide at the base and tapers as it reaches the top – in the same way, the body of the lumbar vertebrae are more bulky and the cervical vertebrae are smaller. Just the opposite, the vertebral foramina must accommodate a relatively larger spinal cord at the top, and a smaller spinal cord at the bottom (Figure 21). Also, from a side view the thoracic vertebrae (makes up the longest section) look like the head of a giraffe (animal with the longest neck).

To remember the atlas and axis, think of Greek mythology and picture Atlas (the first cervical vertebra) holding up the heavens and earth (our head). Then, think of the earth rotating on its axis in the same way we rotate our head when we gesture "no."

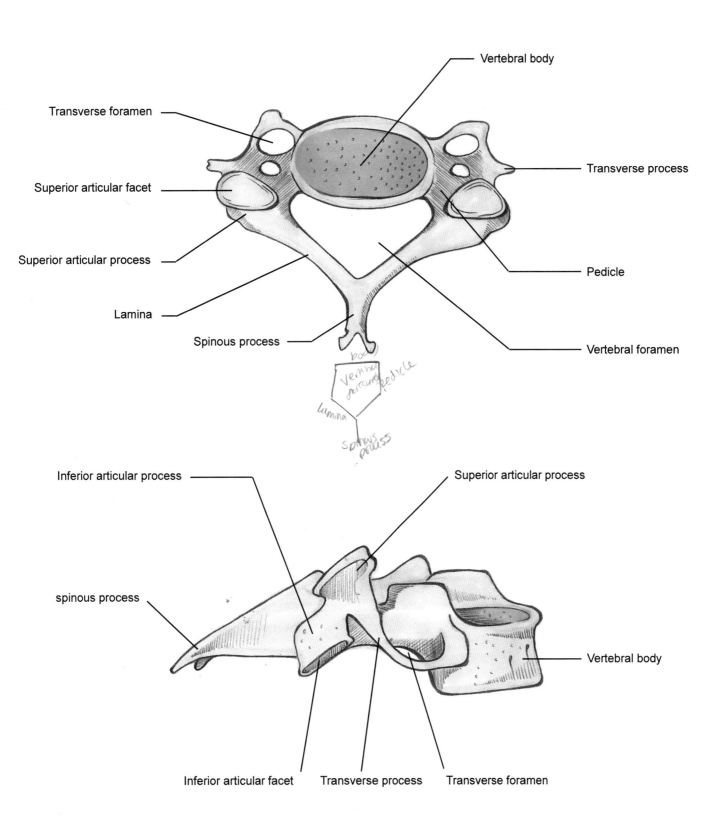

Figure 18. Cervical vertebra. Superior and lateral view.

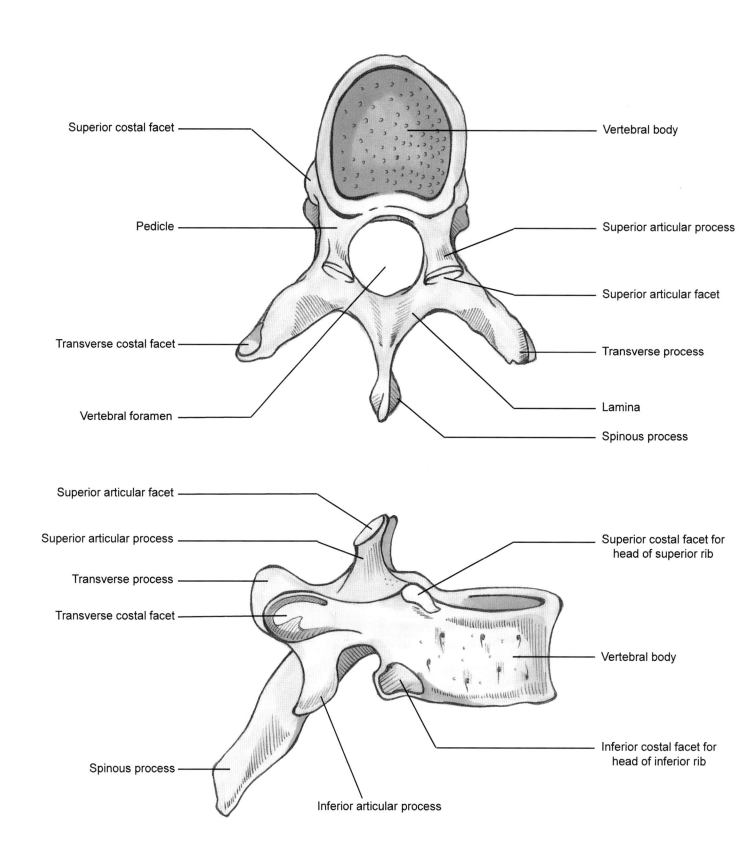

Superior costal facet

Pedicle

Transverse costal facet

Vertebral foramen

Vertebral body

Superior articular process

Superior articular facet

Transverse process

Lamina

Spinous process

Superior articular facet

Superior articular process

Transverse process

Transverse costal facet

Spinous process

Inferior articular process

Superior costal facet for head of superior rib

Vertebral body

Inferior costal facet for head of inferior rib

Figure 19. Thoracic vertebra. Superior and lateral view.

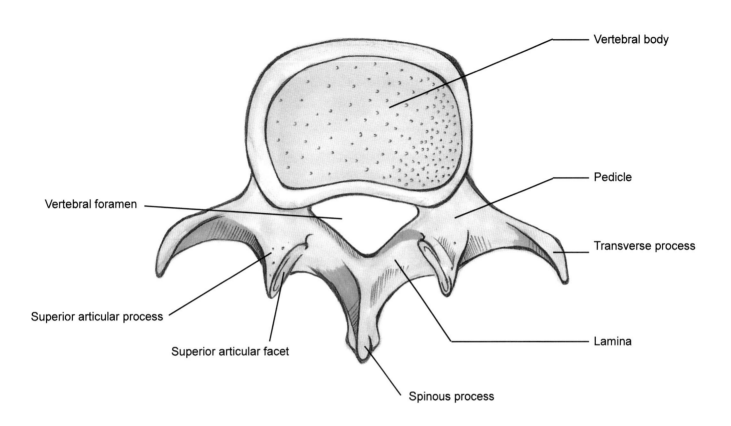

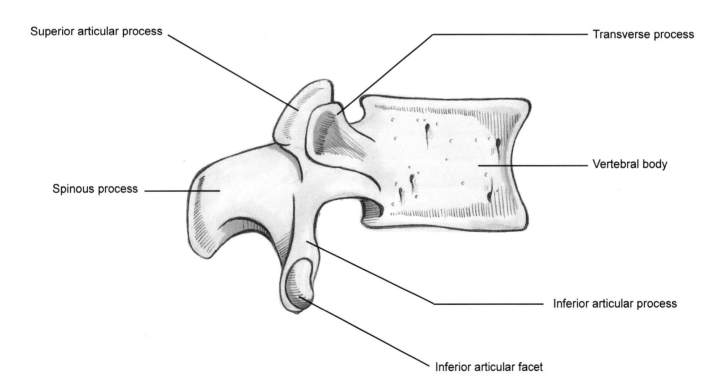

Figure 20. Lumbar vertebra. Superior and lateral view.

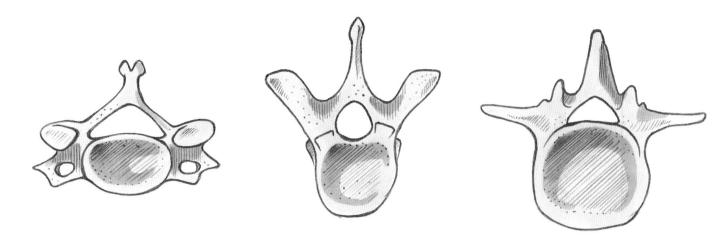

Figure 21. Varying sizes of vertebral bodies and vertebral foramen.

Sacrum

The **sacrum** (*sacrus* = sacred) is wedge-shaped and consists of five sacral vertebrae, which fully fuse together by a person's mid to late twenties; the four **transverse lines** can be seen on the anterior surface (Figure 22). It is a strong bone with large **auricular surfaces** on each side that form the **sacroiliac joint** with the os coxa – this is the formation of the pelvic girdle. On the posterior surface is the **median sacral crest** which is formed by the fusion of the spinous processes, and **sacral foramina** are visible on either side for nerves to pass through from the spinal cord.

Within the sacrum, and continuing down from the vertebral canal, is the **sacral canal**. Its exit point is the **sacral hiatus**. On the posterior side, the **superior articular processes** articulate with L5. The **sacral promontory** is the name given to the superior border on the anterior side of the sacrum.

Coccyx

Articulating on the inferior side of the sacrum is the **coccyx** (*kokkyx* = like a cuckoo's beak), a bone commonly known as our "tailbone." It is triangular in shape and comprised of four or five fused coccygeal vertebrae (Figure 22).

Thoracic cage

he **rib** (**thoracic**) **cage** consists of the thoracic vertebrae, 12 paired ribs, costal cartilage, and the sternum (Figure 23). It serves many functions including: support of the pectoral girdle and upper appendages, housing and protecting organs and the facilitation of breathing.

Ribs

There are 12 pairs of ribs, each one attaching to a thoracic vertebra (Figure 24). The first seven pairs are called **true ribs** or **vertebrosternal ribs** because they affix to the sternum by individual **costal** (*costa* = rib) **cartilages**. The remaining five pairs (ribs 8, 9, 10, 11, and 12) are referred to as **false ribs** because they do not attach to the sternum. Ribs 8, 9, and 10 have costal cartilage that fuses to form the costal margin of the rib cage, for this reason they are called **vertebrochondral ribs** (Figure 23). The last two pairs of false ribs (11 and 12) are also referred to as **floating ribs** because they do not attach to the sternum; rather they are embedded into the muscles of the posterior body wall.

Sternum

The **sternum** (*sternon* = chest) is the breastbone. It is a flat, elongated bone made up of three sections: the **manubrium** (*manubrium* = a handle) is the most superior portion; the **body** is the middle, largest portion; and the **xiphoid** (ZĪ-foyd; *xiphos* = sword-like) **process** makes up the small, inferior portion (Figure 25). At the superior end of the manubrium is the **jugular notch**, with clavicular notches present on either side for articulation with the medial end of the clavicles, thus forming the sternoclavicular joints. The sternocostal joints are also found at the manubrium where it articulates with the costal cartilage of the first and second ribs. The manubrium and body connect at the **sternal angle** (**angle of Louis**; Antoine Louis, French 1723-92). The body of the sternum attaches, directly and indirectly, to the costal cartilages of the second through tenth ribs. The xiphoid process does not attach to any ribs; rather it serves as an attachment point for abdominal muscles.

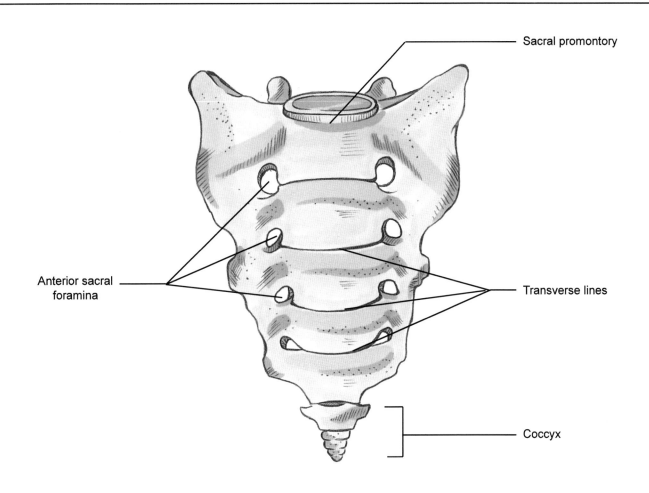

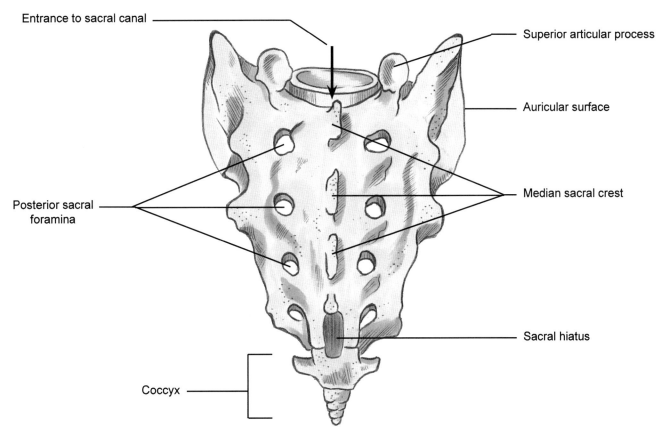

Figure 22. Sacrum and coccyx. Anterior and posterior view.

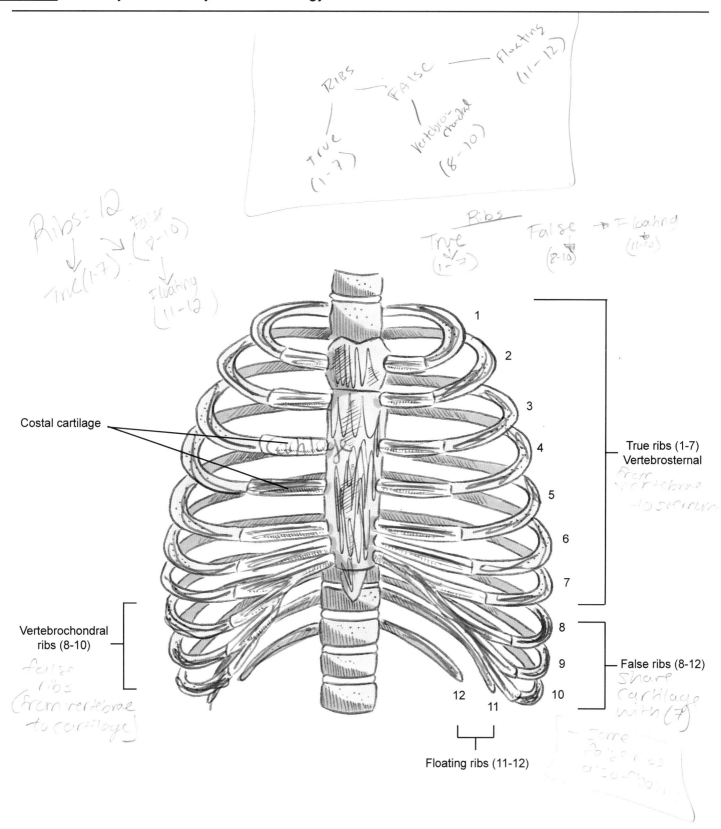

Ribs
FALSE ——— Floating (11-12)
True (1-7)
Vertebrochondral (8-10)

Ribs = 12
True (1-7)   False (8-10)
Floating (11-12)

Ribs
True (1-7)   False (8-10) → Floating (11-12)

Costal cartilage

1
2
3
4
5
6
7
8
9
10
11
12

True ribs (1-7)
Vertebrosternal
from vertebrae to sternum

Vertebrochondral ribs (8-10)
False ribs (from vertebrae to cartilage)

False ribs (8-12)
Share cartilage with (7)
some false ribs also floating

Floating ribs (11-12)

Figure 23. Thoracic cage. Anterior view.

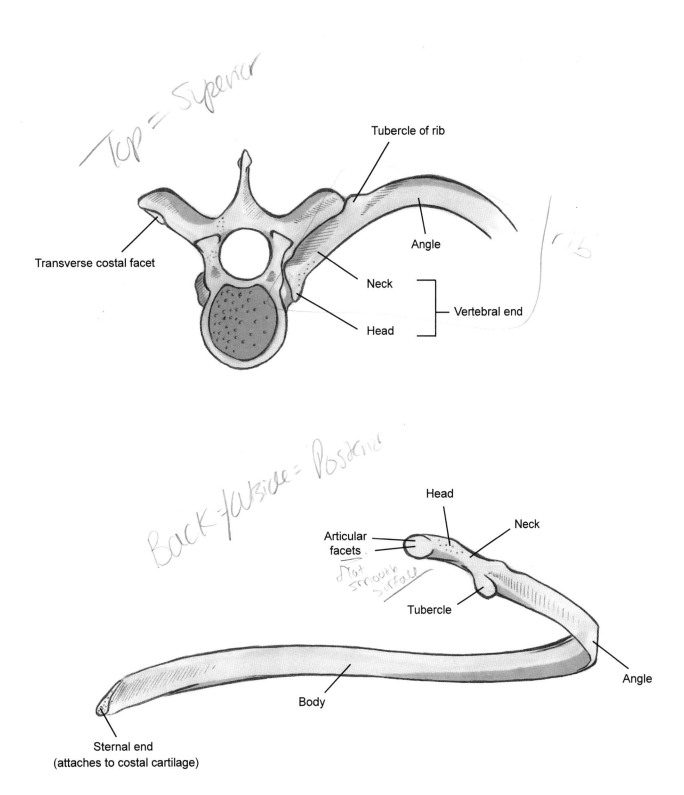

Top = Superior

Back/backside = Posterior

rib

Tubercle of rib

Angle

Transverse costal facet

Neck

Vertebral end

Head

Head

Neck

Articular facets

Not smooth surface

Tubercle

Angle

Body

Sternal end
(attaches to costal cartilage)

Figure 24. Thoracic articulation of rib, superior view. Rib, posterior view.

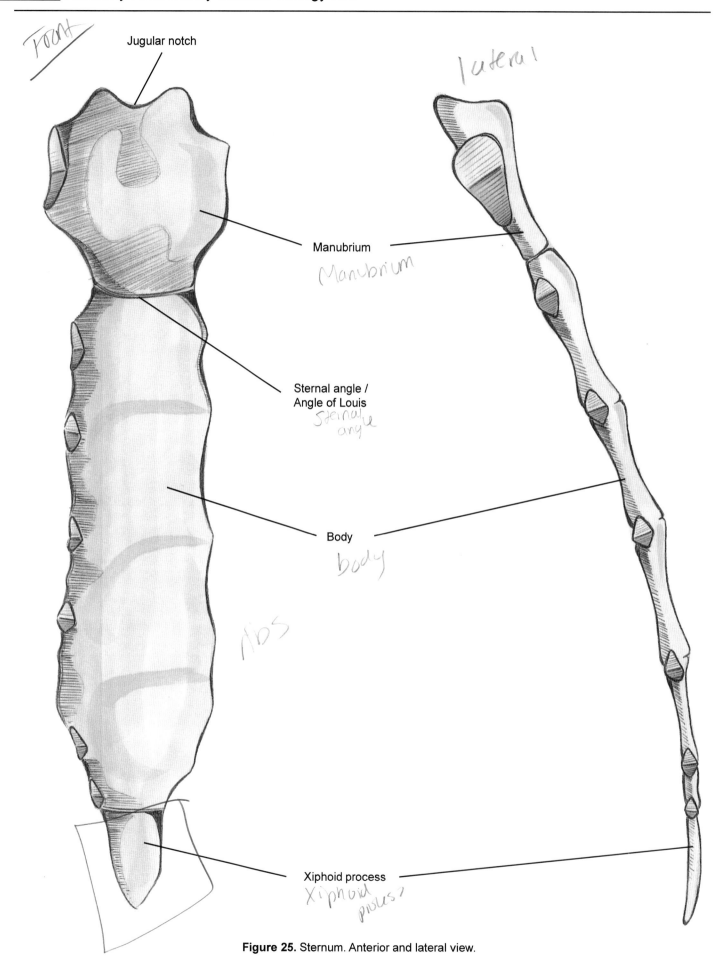

Front

Jugular notch

Manubrium

Manubrium

Sternal angle /
Angle of Louis

Sternal angle

Body

body

Xiphoid process

Xiphoid process?

lateral

ribs

Figure 25. Sternum. Anterior and lateral view.

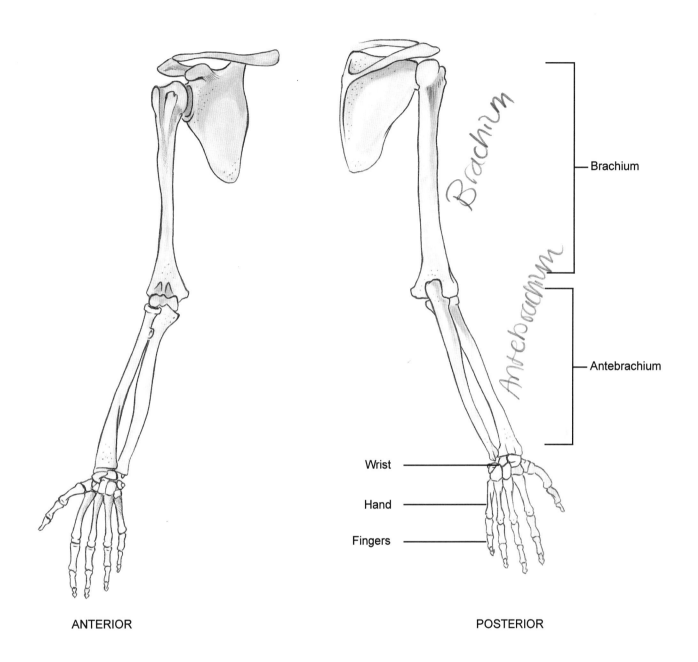

Brachium

Antebrachium

Wrist

Hand

Fingers

ANTERIOR

POSTERIOR

Figure 26. Upper appendicular skeleton. Anterior and posterior view.

Appendicular Skeleton

The term **appendicular** comes from the Latin *appendere* meaning to hang on, or supplement. From here we derive the word appendage, which is defined as an extremity: an external body part that projects from the body; a part that is joined to something larger.

You see? When you break down the words they actually make a lot of sense... blah, blah, blah (if this is boring you, just turn back to Figure 7 for a refresher).

Right

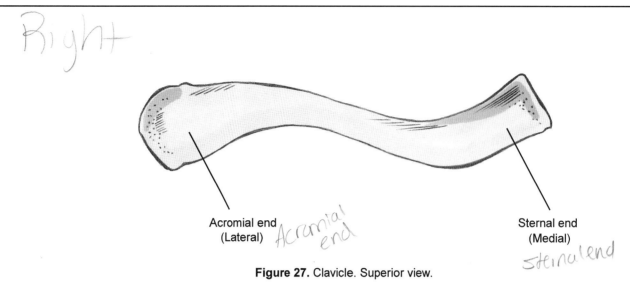

Acromial end
(Lateral)
Acromial end

Sternal end
(Medial)
sternal end

Figure 27. Clavicle. Superior view.

Upper Appendicular Skeleton

The two scapulae and two clavicles make up the **pectoral (shoulder) girdle** – the attachment point for bones of the upper limbs to the axial skeleton (Figure 26). The **sternoclavicular joint** is the only direct bony connection the entire arm has with the axial skeleton. The scapula is held in position by several muscle attachments, the **acromioclavicular joint**, and the **glenohumeral (shoulder) joint**. The pectoral girdles do not articulate with vertebrae and are not weight-bearing, thus making them freely movable yet somewhat delicate.

Clavicle
The clavicles lie superior to the first rib and serve as the only bony attachment of the upper appendages to the axial skeleton. Each **clavicle** (*clavicula* = small key) or collarbone, is an S-shaped bone with a **sternal end** on the medial extremity (sternoclavicular joint) and an **acromial end** on the lateral extremity (acromioclavicular joint) (Figure 27).

Scapula
Each **scapula** (shoulder blade) is a large, triangular, flat bone found on the posterior side of the rib cage, located between the second and seventh ribs (Figure 28). The **spine** is a ridge that diagonally crosses the posterior side, creating a **supraspinous fossa** above and an **infraspinous fossa** below. The spine widens as it runs laterally to the **acromion** (*akros* = peak, summit; *omos* = shoulder). Inferior to the acromion is the **glenoid cavity**, a shallow depression in which the head of the humerus fits (glenohumeral joint). Superior and anterior to the glenoid cavity is an upward projection called the **coracoid** (*korakodes* = crow's beak) **process** to which muscles attach. The anterior surface of the scapula is termed the **subscapular fossa**.

The edges of the scapula are easy to remember – the **superior border** on top, the **medial (vertebral) border** closest to the midline/vertebral column, and the **lateral border** which directs out to the arm. The **superior angle** is located between the superior and medial borders. The **inferior angle** is between the medial and lateral borders. Along the superior border is a depression called the **scapular notch**, an area that accommodates nerve supply.

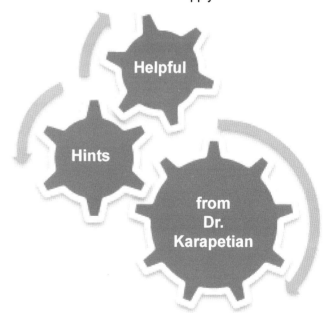

To determine if you are looking at a right or left scapula (without touching it), first identify whether or not its spine is visible. If it is, you are looking at the posterior side; if it is not, you are looking at the anterior side. From either position, next imagine the arm attaching from the torso laterally from the glenohumeral joint.

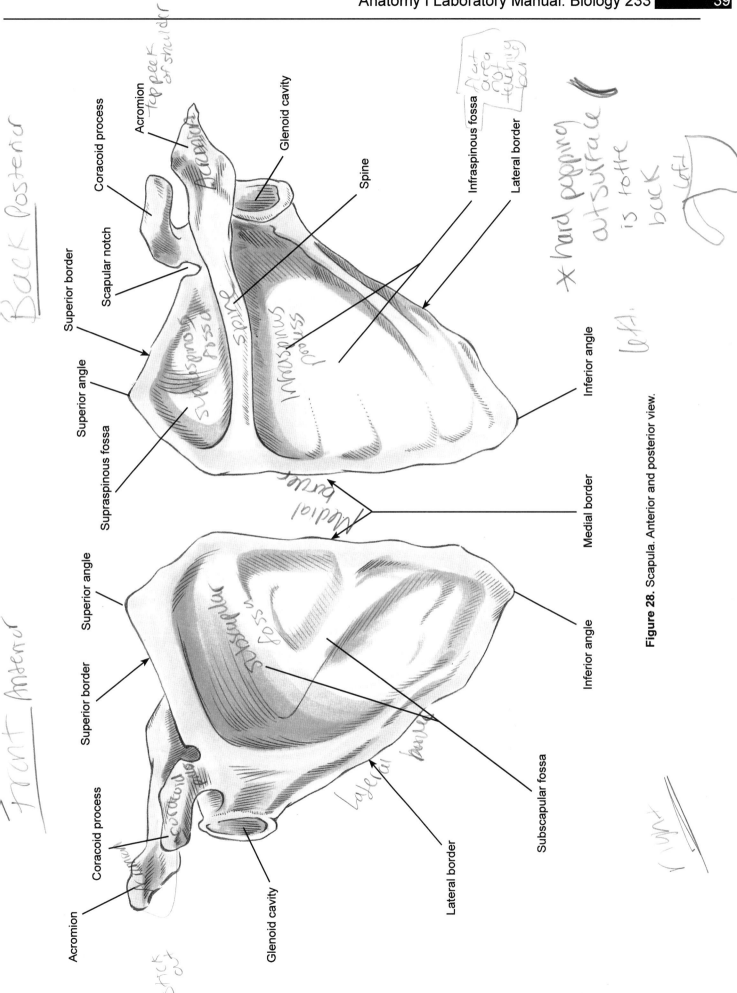

Figure 28. Scapula. Anterior and posterior view.

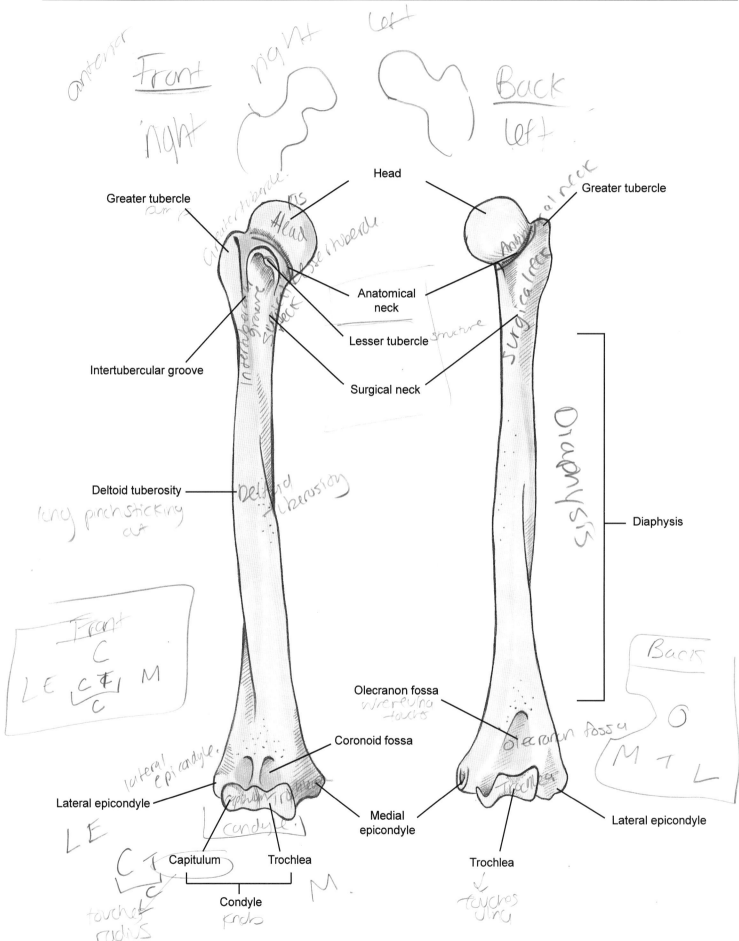

Figure 29. Humerus. Anterior and posterior view.

Humerus

The term **brachium** refers to the upper arm and consists of just one bone, the **humerus** – the largest and longest bone of the upper limb (Figure 29). It has a proximal **head** that articulates with the glenoid cavity of the scapula, a **diaphysis** (dī-AF-uh-sis; *dia* = through; *physis* = growth) or **shaft** that makes up the length of the bone, and a distal end (elbow) that meets with the two bones of the forearm. The **anatomical neck** is the indented groove that surrounds the head. The **surgical neck** is the area where the shaft begins to narrow. Two eminences are evident at the proximal end of the humerus: the **greater tubercule** and the **lesser tubercule**; these are separated by the **intertubercular (bicipital) groove**, a shallow passageway that accommodates the tendon of the biceps brachii.

The **deltoid tuberosity** is a roughened area midway down the shaft on the lateral side; it is the site where the deltoid muscle attaches to the bone. The distal end includes the **capitulum** which articulates with the radius, and the **trochlea** (trōk' le-uh; *trochilia* = a pully) which articulates with the ulna. On either side of these condylar surfaces are the **medial** and **lateral epicondyles**. On the anterior side, the **coronoid** (*korne* = crown-shaped) **fossa** is a visible indentation above the trochlea. On the posterior side, the **olecranon fossa** is a large depression. These fossae create receiving ends for the ulna when we bend our elbows.

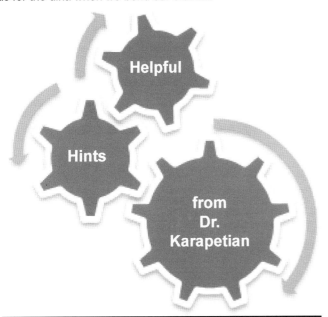

Helpful Hints from Dr. Karapetian

To determine if you are looking at a right or left humerus (without touching it), first identify whether or not you can see the intertubercular groove. If you can, you are looking at the anterior side of the bone; if you cannot, you are looking at the posterior (and will therefore see the olecranon fossa). Next, picture how the head of the humerus would have to face medially in order to fit into the glenoid cavity of the scapula.

Antebrachium (Forearm)

The ulna and radius make up the forearm. The ulna has a more complete connection to the humerus than does the radius, however the radius makes a larger contribution to the wrist.

Radius

The **radius** has a small proximal end and a large distal end. The proximal **head** looks like the radius of a circle which articulates with the radial notch of the ulna and the capitulum of the humerus. The **radial tuberosity** serves as the attachment point for the biceps brachii muscle. The relatively large distal end has an **ulnar notch** and **styloid process** (Figure 30).

Ulna

The proximal end of the **ulna** has a semilunar structure called the **trochlear notch** which articulates with the trochlea of the humerus. The **olecranon** forms the posterior of this trochlear notch, thus creating the elbow. The **coronoid process** forms the anterior lip of the trochlear notch. The **radial notch** allows for the head of the radius to fit in place. At the distal end are the **head** and **styloid process** (Figure 30). The styloid processes of the radius and ulna offer lateral and medial support of the wrist, respectively.

Hand

There are 27 bones of the hand, divided into three regions: proximal carpus, intermediate metacarpus, and distal phalanges (Figure 31).

Carpus

The **carpus** (*karpos* = wrist) is comprised of eight **carpal bones** residing in two rows of four bones. The proximal row, from lateral (thumb) to medial, is the **scaphoid** (*skaphe* = boat), **lunate** (*luna* = moon), **triangular** (or **triquetrum**) (*triangulum* = three angled; *triquetrus* = three cornered), and **pisiform** (*pisum* = pea shaped). The distal row, from lateral to medial, consists of the **trapezium** (*trapezion* = four-sided with no parallel sides), **trapezoid** (*trapezoids* = trapezoid shaped), **capitate** (*caput* = head), and **hamate** (*hamatus* = hook).

Metacarpus

The metacarpus is the palm of our hand and is made up of five **metacarpal bones**. Each bone has a base, shaft, and distal head which form our knuckles when we clench our fists. These bones are numbered, lateral to medial, from I to V, the thumb being I.

Phalanges

There are 14 phalanges that make up our digits. A single bone of a finger is called a **phalanx** (*phalanx* = closely knit row). Like the metacarpals, the phalanges are numbered I to V, beginning with the thumb. The phalanges of the fingers are arranged in proximal, middle, and distal rows – however the thumb (pollex) only has a proximal and distal phalanx.

Radus/ulna

right

Radius   Top→Bottom

Radial head
Radial tuberosity
Styloid process

Ulna

ulna

right

Ulna

Olecranon
Trochlear notch
Coronoid process
Radial notch
Diaphysis
Ulna notch
Ulna head
Styloid proc

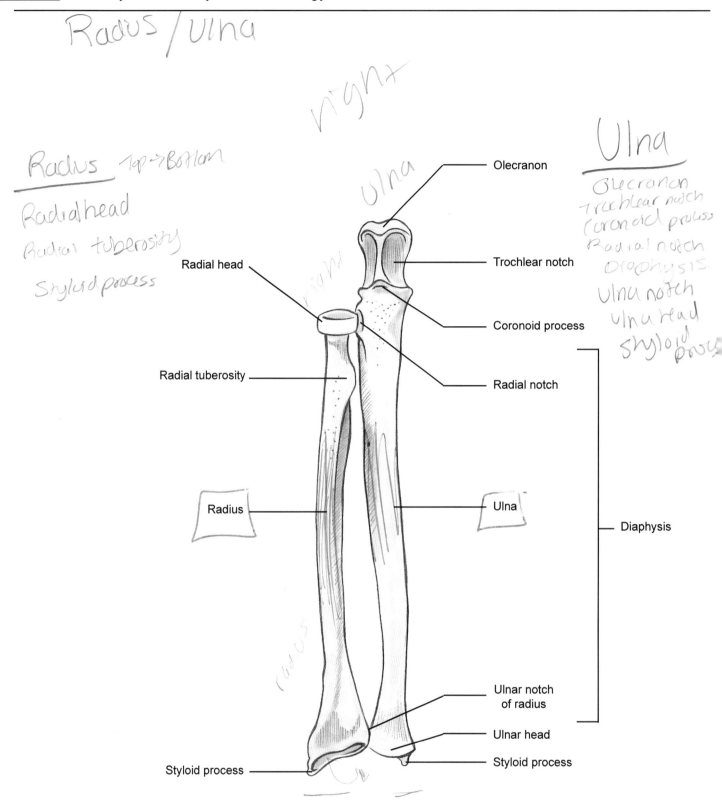

Radial head

Radial tuberosity

Radius

radius

Styloid process

Olecranon

Trochlear notch

Coronoid process

Radial notch

Ulna

Ulnar notch
of radius

Ulnar head

Styloid process

Diaphysis

Figure 30. Radius and ulna. Anterior view.

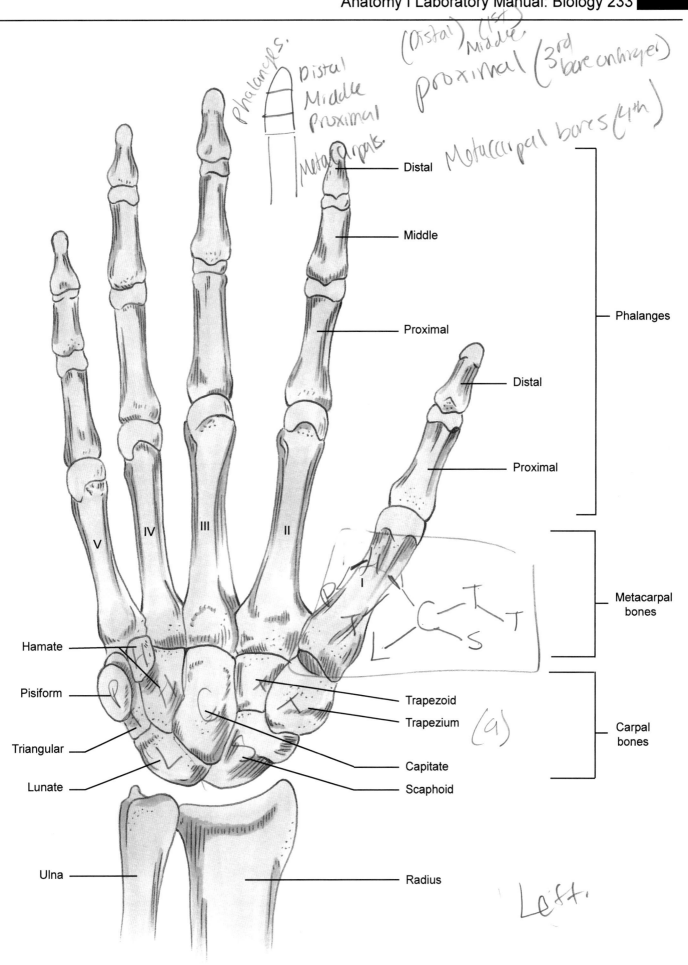

Phalanges.
Distal
Middle
Proximal
Metacarpals.

(Distal) Middle
Proximal (3rd bone cartilage)
Metacarpal bones (4th)

Distal

Middle

Proximal

Distal

Proximal

Phalanges

Metacarpal bones

Carpal bones

V   IV   III   II

CTTS

(9)

Hamate

Pisiform

Triangular

Lunate

Trapezoid

Trapezium

Capitate

Scaphoid

Ulna

Radius

Left.

Figure 31. Hand and wrist. Anterior view.

Lower Appendicular Skeleton

he weight of our upper body is supported into the floor by the lower appendicular skeleton. Our pelvis consists of three bones fused together which attach to the femur (our thigh). The patella serves as our kneecap. The bones of the leg, the tibia and fibula, then continue downward to articulate with the bones of the ankle, which is followed by the bones of the foot and toes.

Pelvic (Hip) Girdle

The **pelvic girdle** (**pelvis**) is formed by the **ossa coxae**, the paired **hip bones** (Figure 32). These bones are united by the **pubic symphysis** on the anterior side, and the sacrum on the posterior side (Figure 33). The pelvic girdle serves several functions including supporting the weight of the body from the vertebral column, and protecting the lower viscera, i.e.: the urinary bladder, the internal reproductive organs, and a developing fetus in pregnant women. The pelvis can be divided into a **greater** or **false pelvis** and a **lesser** or **true pelvis**. A curved bony frame called the **pelvic brim** separates these two regions (Figure 34). The pelvic brim surrounds the **pelvic outlet**, where the greater pelvis lies superior, and during parturition (childbirth), a baby passes through the mother's lesser pelvis for a natural delivery... and holy smokes, that must hurt like heck!

Each **os coxae** (singular **hip bone**; os - COCKS - ē; plural **ossa coxae** is the paired hip bones or pelvic girdle) is comprised of three separate bones: the **ilium**, the **ischium**, and the **pubis**. On the lateral surface of the adult hip bone, these three bones are fused together creating a large circular depression called the **acetabulum** (*acetabulum* = vinegar cup), which houses the head of the femur.

Ilium

The **ilium** (*ilia* = loin) is the largest of the three pelvic bones. The **iliac crest** is the uppermost landmark of the hips and terminates anteriorly as the **anterior superior iliac spine**. Just below this is the **anterior inferior iliac spine**. The posterior end of the iliac crest is the **posterior superior iliac spine**, and just below this is the **posterior inferior iliac spine**.

elow the posterior inferior iliac spine is the **greater sciatic** (sī-AT-ik) **notch**. On the medial surface is the **auricular surface**, a roughened area where the sacrum articulates with the hips. On the anterior is the **iliac fossa**, a smooth, concave area from where the iliacus muscle originates. The sacroiliac ligament attaches at the **iliac tuberosity** which is posterior to the iliac fossa. On the gluteal surface, three coarsened ridges are present: the inferior, anterior, and posterior gluteal lines from which the gluteal muscles attach.

Ischium

The **ischium** (ISS-kē-um; *ischion* = hip joint) makes up the posterior and inferior component of the os coxae. Directly inferior and posterior to the greater sciatic notch is the spine of the ischium. Inferior to this spine is the **lesser sciatic notch**. The **ischial tuberosity** is the bony projection that supports our body weight when we sit down. The **obturator foramen**, used for muscle attachment, is formed by the ramus of the ischium, along with the pubis.

Pubis

The **pubis** (*pubis* = genital area) is the anterior bone of the os coxae. It contains a **superior ramus** and **inferior ramus** that are on either side of the body of the pubis. The body is what helps form the pubic symphysis.

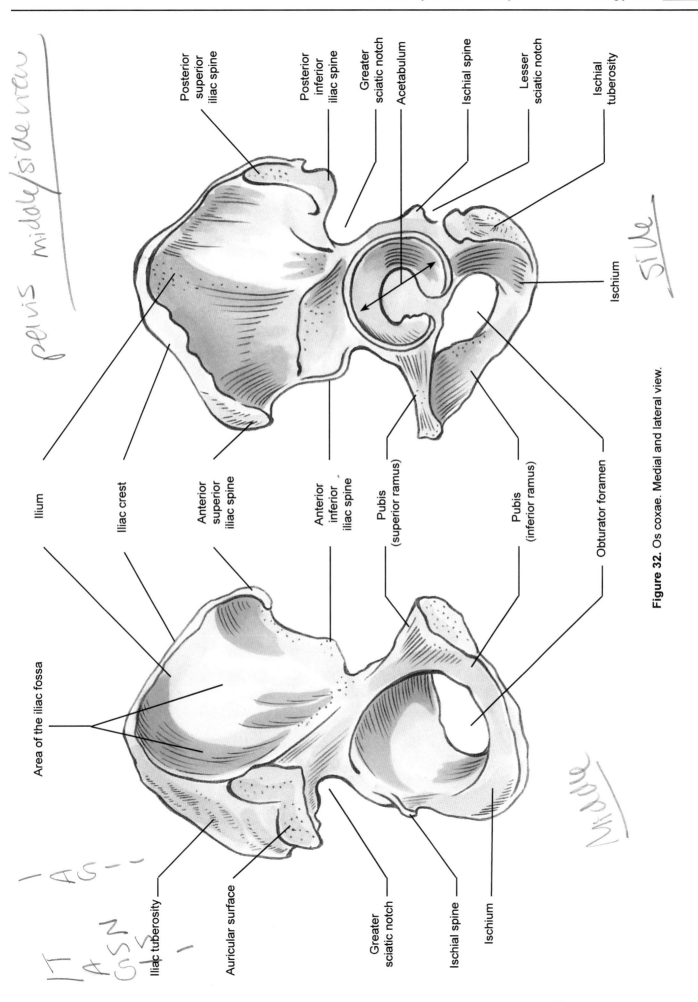

Figure 32. Os coxae. Medial and lateral view.

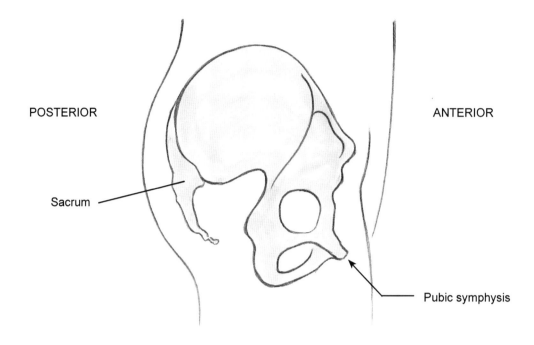

POSTERIOR

ANTERIOR

Sacrum

Pubic symphysis

Figure 33. Position of the os coxae. Lateral view.

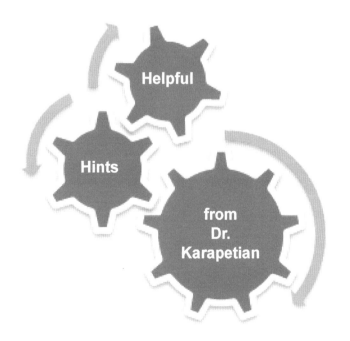

Helpful

Hints

from Dr. Karapetian

Alright, let's cut to the chase: learning the hips can be pretty difficult. There are a ton of new terms, and you're probably going to have to point out if it's a left or right os coxae on an exam. To determine if you're looking at a right or left hip, be able to visualize where the hip joint extends out to your leg. Remember the pubic symphysis and the acetabulum? No? Well you'd better, because they're going to help you out in winning this battle. Picture the pubic symphysis anterior (which makes the auricular surface posterior) and the acetabulum facing outward – that's "lateral" in anatomical speak. This makes sense because our hip joints and legs face outward from our midline, not inward. Hold the individual hip bone models next to you – it will help.

Differences in the Male and Female Pelvis

tructural differences between the adult male and female pelvis largely reflect the role this area plays during pregnancy and parturition (childbirth) in the female (Figure 34). Aside from bony anatomical differences (Table 7), the pubic symphysis and sacroiliac joints have the ability to stretch during pregnancy.

Femur
The **femur** (*femur* = thigh) is the strongest and longest bone in the body (Figure 35). At the proximal end is a round **head** that articulates with the acetabulum. The **fovea capitis** is a shallow divot in the middle of the head where the ligamentum teres holds this attachment between the head and the acetabulum securely.

he narrow region beneath the head is the **neck**, followed by a lateral **greater trochanter** and a medial **lesser trochanter**. The **diaphysis** (**shaft** or **body**) of the femur bows slightly to the midline, with the long **linea aspera** (*linea* = line; *asperare* = rough) visible on its posterior surface. The distal end expands for articulation with the tibia at the **medial** and **lateral condyles**. On the posterior side the depression between the condyles is the **intercondylar fossa**, and the area superior to that is the **popliteal surface**. The **patellar surface** is the area between the condyles on the anterior side. The **medial** and **lateral epicondyles** are the areas above the respective medial and lateral condyles, and serve as ligament and tendon attachment sites.

Table 7. Comparison of the male and female pelvic girdle

CHARACTERISTIC	MALE PELVIS	FEMALE PELVIS
General structure	More massive; prominent processes	More delicate; processes less prominent
Pelvic inlet	Heart shaped	Round or oval shaped
Pelvic outlet	Narrow	Wide
Anterior superior iliac spines	Not very wide apart	Wide apart
Obturator foramen	Oval shaped	Triangular shaped
Pubic symphysis	Deep, long	Shallow, short
Pubic arch	Less than 90°	Greater than 100°
Acetabulum	Faces laterally	Faces more anteriorly

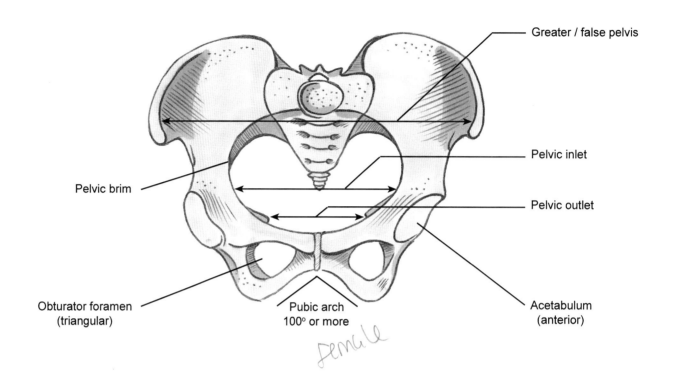

Greater / false pelvis

Pelvic inlet

Pelvic outlet

Pelvic brim

Acetabulum
(anterior)

Obturator foramen
(triangular)

Pubic arch
100° or more

female

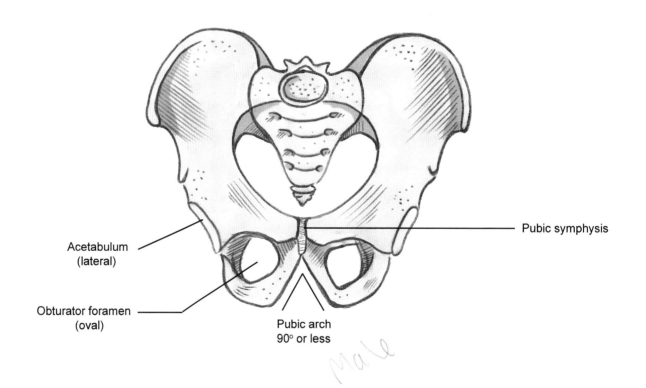

Pubic symphysis

Acetabulum
(lateral)

Obturator foramen
(oval)

Pubic arch
90° or less

male

Figure 34. Comparison of female and male pelvis. Anterior view.

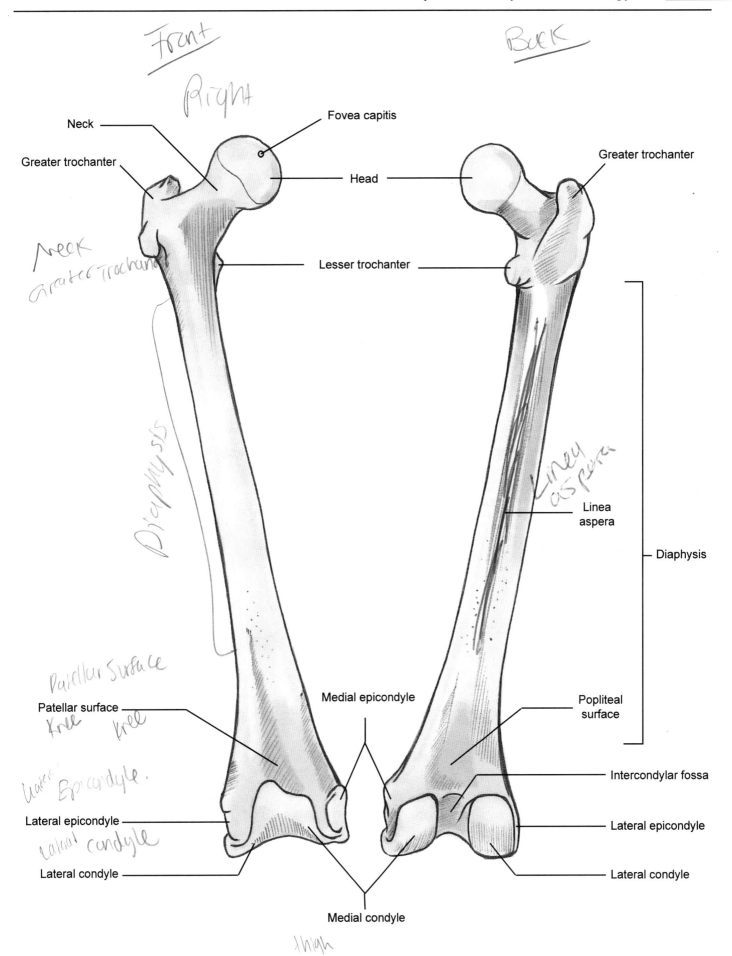

Figure 35. Femur. Anterior and posterior view.

Patella

he kneecap is properly termed the **patella** (*patina* = small plate). It is the triangular sesamoid bone found on the anterior side of the knee joint, and has a broad **base** and inferiorly pointed **apex**. **Articular facets** on the posterior surface articulate with the medial and lateral condyles of the femur.

Tibia

The **tibia** (*tibia* = pipe, flute) is commonly called the shinbone (Figure 36). It articulates with the femur at its proximal end to bear our bodyweight, and the talus as the distal end to form the ankle. There are two concave surfaces on the proximal end, the **medial** and **lateral condyles**, which articulate with the condyles of the femur. The **tibial tuberosity** is located on the proximoanterior part of the tibia, and serves as an attachment point for the patellar ligament. If you palpate your knee while it is bent at 90°, the tibial tuberosity is the "bump" you feel just inferior to your patella. The **anterior crest (border)** is the sharp ridge that runs the length of the shaft of the bone. The **medial malleolus** is the medial bony protrusion located on the medial side of the distal end of the bone, this is easily palpated as the "inside bump" of your ankle.

Fibula

The **fibula** is more important for muscle attachment than it is for weight support; this is evidenced as you look at the long, narrow bone (Figure 36). The **head** articulates with the proximolateral end of the tibia. The **lateral malleolus** is the bony knob at the distal end; this is easily palpated as the "outside bump" of your ankle.

Foot

The tarsus, metatarsus, and phalanges make up the 26 bones of the foot (Figure 37). Their layout is similar to that of the hand, but structurally they must be able to withstand our bodyweight and provide leverage when we walk.

here are seven **tarsal** (*tarsos* = flat of the foot) bones. The **talus** (*talus* = ankle) articulates with the tibia; the **calcaneus** (*calcis* = heel) is the largest tarsal bone and is the heel of the foot; the **navicular** bone is anterior to the talus; the **medial**, **intermediate**, and **lateral cuneiform** bones along with the **cuboid bone** form the anterior line of tarsal bones from the medial to lateral side, respectively.

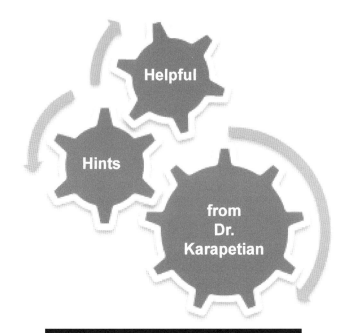

Remembering bones of the ankle can be tricky, but try to memorize this easy mnemonic: Tom Can Control Not Much In Life. Also, to help keep some of the ankle bones in order – think "TTT" for Talus Top Tibia, because the talus is the top of the ankle bones and articulates with the tibia. Finally, remember the calcaneus is the heel of the foot where the calcaneal (Achilles) tendon attaches.

There are five **metatarsal** bones numbered I to V from medial to lateral. The 14 **phalanges** make up the toes and as in the hand, are arranged by proximal, middle, and distal rows. Similar to the thumb of the hand, the great toe (hallux) only has a proximal and distal phalanx.

wo arches support the weight of the body and provide leverage while walking. They are the **longitudinal arch** which runs the length of the foot, and the **transverse arch** which extends across the width of the foot. These arches are formed by the arrangement of the bones and are held in place by ligaments and tendons. When these ligaments and tendons weaken, the height of the longitudinal arch decreases – leading to a condition called *pes planus*, or flatfoot. Interestingly, a persistent case of flatfoot may prevent someone from successfully enlisting in the US Army... can you imagine marching miles and miles without proper arch support? Ouch!

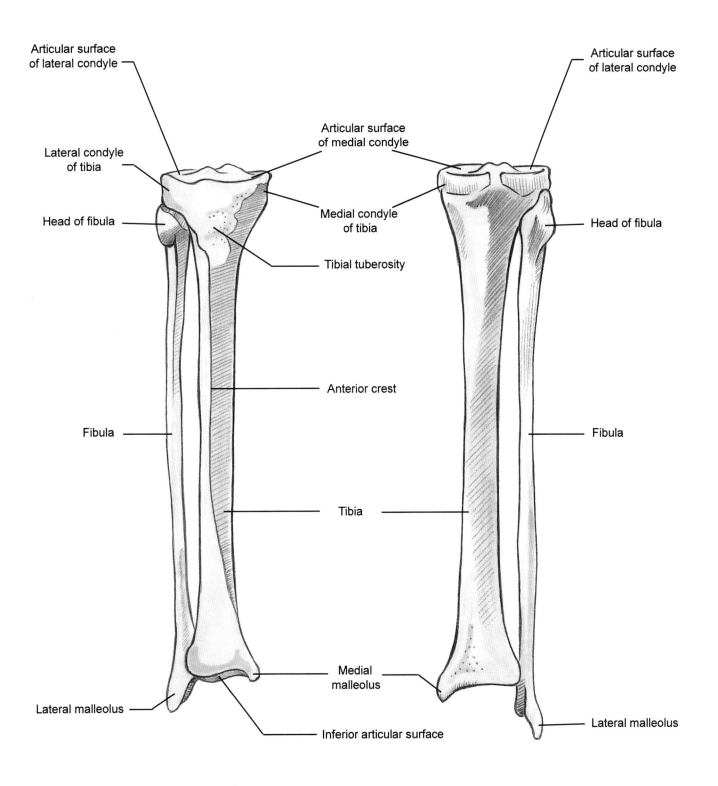

Figure 36. Tibia and fibula. Anterior and posterior view.

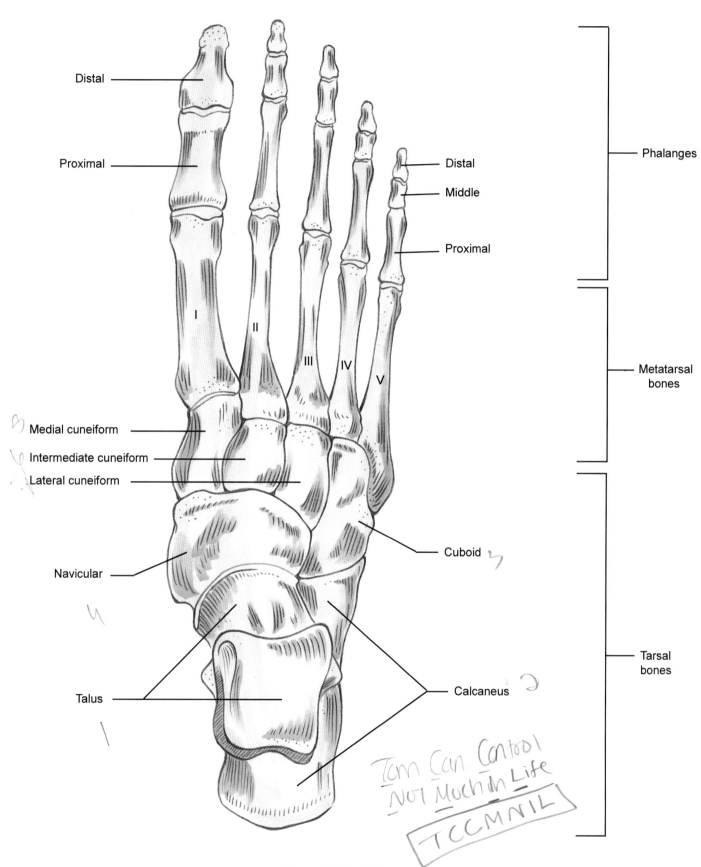

Distal

Proximal

Distal

Middle

Proximal

I

II

III

IV

V

Medial cuneiform

Intermediate cuneiform

Lateral cuneiform

Navicular

Talus

Cuboid

Calcaneus

Phalanges

Metatarsal bones

Tarsal bones

Tom Can Control
Not Much in Life
TCCMNIL

Figure 37. Foot. Superior view.

UNIT II

Introduction

The term **muscle tissue** is used in reference to all the contractile tissues of the human body – this includes cardiac, smooth, and skeletal muscle. Cardiac muscle is located in the heart and is therefore considered part of the cardiovascular system. Smooth muscle tissue can be found in the intestines, thereby part of the digestive system; or in the urinary bladder, thus part of the urinary system. Skeletal muscle is found throughout the human body so when we use the term **muscular system**, we are referring to the voluntary skeletal muscles (Figure 38 and 39).

In this next unit we are going to study the voluntary skeletal muscles. But before we do, there are 1000 new terms you will be responsible for knowing. Just kidding… there are only about 40 terms. Now you're probably thinking to yourself, "Forty terms before we even learn the muscles?!?" Yes aspiring anatomy students, 40. Well, actually 39… see, things are looking better for you already! And they are a pretty easy 39; most of them make good sense when you read the definition. Plus, knowing these terms will make learning the actual muscles much easier. See Tables 8 – 11 below.

Nomenclature

Table 8. Characteristics used to name muscles: Action

TERM	MEANING	EXAMPLE
Flexor	Decreases joint angle	Flexor carpi radialis
Extensor	Increases joint angle	Exensor carpi radialis
Abductor	Moves bone away from midline	Abductor pollicis longus
Adductor	Moves bone closer to midline	Adductor magnus
Supinator	Turns palm superiorly or anteriorly	Supinator
Pronator	Turns palm inferiorly or posteriorly	Pronator teres
Levator	Produces superior movement	Levator scapulae
Depressor	Produces inferior movement	Depressor anguli oris
Tensor	Makes a body part rigid	Tensor fasciae latae
Sphincter	Decreases size of opening	External anal sphincter

Table 9. Characteristics used to name muscles: Size

TERM	MEANING	EXAMPLE
Longus	Long	Fibularis longus
Longissimus	Longest	Longissimus capitis
Teres	Long and round	Teres major
Brevis	Short	Fibularis brevis
Magnus	Large	Adductor magnus
Major	Larger	Pectoralis major
Maximus	Largest	Gluteus maximus
Minor	Small	Pectoralis minor
Minimus	Smallest	Gluteus minimus
Vastus	Great	Vastus lateralis

Table 10. Angular movements

TERM	DESCRIPTION
Extension	Involves an increase in the angle between articulating bones; usually in the sagittal plane
Hyperextension	Continuation of extension beyond the anatomical position
Flexion	Involves a decrease in the angle between articulating bones; usually in the sagittal plane
Lateral flexion	Movement of the torso in the frontal plane
Abduction	Movement of a bone away from the midline; usually in the frontal plane
Adduction	Movement of a bone toward the midline; usually in the frontal plane
Circumduction	A combination of flexion, abduction, extension, and adduction in succession, in which the distal end of a part of the body describes a circle

Table 11. Special movements – Occur at specific joints

TERM	DESCRIPTION
Elevation	Movement of a part of the body superiorly
Depression	Movement of a part of the body inferiorly
Supination	Movement of the forearm in which the palm is turned anteriorly or superiorly
Pronation	Movement of the forearm in which the palm is turned posteriorly or inferiorly
Inversion	Movement of the soles of the foot medially so they face toward each other
Eversion	Movement of the soles of the foot laterally so they face away from each other
Dorsiflexion	Bending the foot in the direction of the dorsum (superior surface)
Plantar flexion	Bending the foot in the direction of the plantar surface (sole)
Protraction	Movement of a part of the body to the anterior; in the transverse plane
Retraction	Movement of a part of the body to the posterior; in the transverse plane
Opposition	Movement of the thumb across the palm to touch the tips of the fingers on the same hand
Rotation	Movement of a bone around its own longitudinal axis; in the limbs, it may be medial (toward the midline) or lateral (away from the midline)

eep in mind that bending and moving are muscular actions but the muscles involved do not typically work alone, rather they work in groups. The coordinated actions of the muscles help to make our movements more efficient, and based on their function each of the muscles involved can be described as an **agonist**, **antagonist**, or **synergist**.

Agonist – the muscle that is the **prime mover**, and therefore doing most of the work, in a specific movement. Think of what your biceps brachii does when you flex or "bend" your elbow.

Antagonist – the muscle that opposes the action of the agonist. Think of what your triceps brachii does when you make that same "bending" movement at your elbow; it helps to assure a smooth movement occurs.

Synergist – (syn = together; ergon = work) the muscle that assists a larger agonist muscle. Synergist muscles can help to stabilize the origin of the agonist (we call these fixators) or provide more pull near the insertion.

Muscles by Region

As you study the following lists, read each individual word carefully – DO NOT be intimidated by big words. In most cases, the name of the muscle actually tells you where its location is, or how it is setup. Also, in the origin and insertion charts at the end of this unit you will see the breakdown of the muscle names to help you learn them more easily... Please, there is no need to thank me (insert sarcasm here).

Muscles of the head (15)
(Figure 40 and 41)

Frontal belly of occipitofrontalis (Frontalis)
Occipital belly of occipitofrontalis (Occipitalis)
Temporoparietalis
Temporalis
Orbicularis oculi: (or-BIC-you-laris OC-you-lie)
    Orbital part
    Palpebral part
Nasalis
Levator labii superioris
Zygomaticus minor
Zygomaticus major
Orbicularis oris
Depressor labii inferioris
Mentalis
Buccinator (BUCKS-in-ā-tor)
Masseter

Notable structure (1)
Epicranial aponeurosis

Muscles of the neck (7)
(Figure 40 and 41)

Sternocleidomastoid
Splenius capitis
Levator scapulae
Scalenes:
    Anterior
    Middle
    Posterior
Platysma (plah-TIS-muh)

Muscles of the torso (14)
(Figure 38, 39, 42, 43)

Trapezius
Latissimus dorsi
Serratus anterior (sir-AY-tiss)
Rhomboid minor
Rhomboid major
Pectoralis major
Pectoralis minor
External intercostals
Internal intercostals
Diaphragm
Rectus abdominis
External oblique
Internal oblique
Transversus abdominis

Notable structures (3)
Thoracolumbar fascia
Linea alba
Rectus sheath

Muscles of the shoulder girdle (8)
(Figure 43 and 44)

Supraspinatus
Infraspinatus
Teres minor (TEAR-eez)
Teres major
Latissimus dorsi
Subscapularis
Deltoid
Pectoralis major

Muscles of the brachium (Arm)
(Figure 45, 46, 47) (7)

Biceps brachii:
    Short head
    Long head
Brachialis
Triceps brachii:
    Lateral head
    Long head
    Medial head
Coracobrachialis

Notable structures (2)
Biceps brachii tendon
Bicipital aponeurosis

Muscles of the antebrachium (Forearm) (Figure 45, 46, 47) (14)

Pronator teres
Flexor carpi radialis
Palmaris longus
Flexor digitorum superficialis
Flexor carpi ulnaris
Extensor carpi ulnaris
Extensor digitorum
Extensor carpi radialis brevis
Extensor carpi radialis longus
Brachioradialis
Abductor pollicis longus (PAUL-ih-sis LONG-us)
Extensor pollicis brevis (PAUL-ih-sis BREV-iss)
Anconeus
Supinator

Notable structures
Flexor retinaculum
Extensor retinaculum (2)

Muscles of the hand
(5)

Flexor pollicis brevis
Abductor pollicis brevis
Adductor pollicis
Flexor digiti minimi
Abductor digiti minimi

Muscles of the hip and thigh
(Figure 48, 49, 50, 51) (19)

Gluteal group:
    Gluteus maximus
    Gluteus medius
    Gluteus minimus
Iliopsoas:
    Psoas major (SO-azz Mā-jer)
    Iliacus
Pectineus
Tensor fasciae latae (Ten-sir FASH-ē-ē LAY-tee)
Sartorius
Adductor group:
    Adductor longus
    Adductor magnus
    Adductor brevis
Gracilis
Quadricep group:
    Rectus femoris
    Vastus medialis
    Vastus lateralis
    Vastus intermedius
Hamstring group:
    Biceps femoris
    Semitendinosus
    Semimembranosus

Notable structure (1)
Iliotibial tract

Muscles of the leg
(Figure 48, 49, 50, 51) (10)

Tibialis anterior
Extensor digitorum longus
Fibularis (Peroneus) longus
Fibularis (Peroneus) brevis
Flexor digitorum longus
Flexor hallucis longus (ha-LOO-sis)
Gastrocnemius: (Gas-trok-Nee-mee-us)
    Medial head
    Lateral head
Soleus (So-lee-us)

Notable structures (4)
Patellar ligament
Calcaneal (Achilles) tendon
Superior extensor retinaculum
Inferior extensor retinaculum

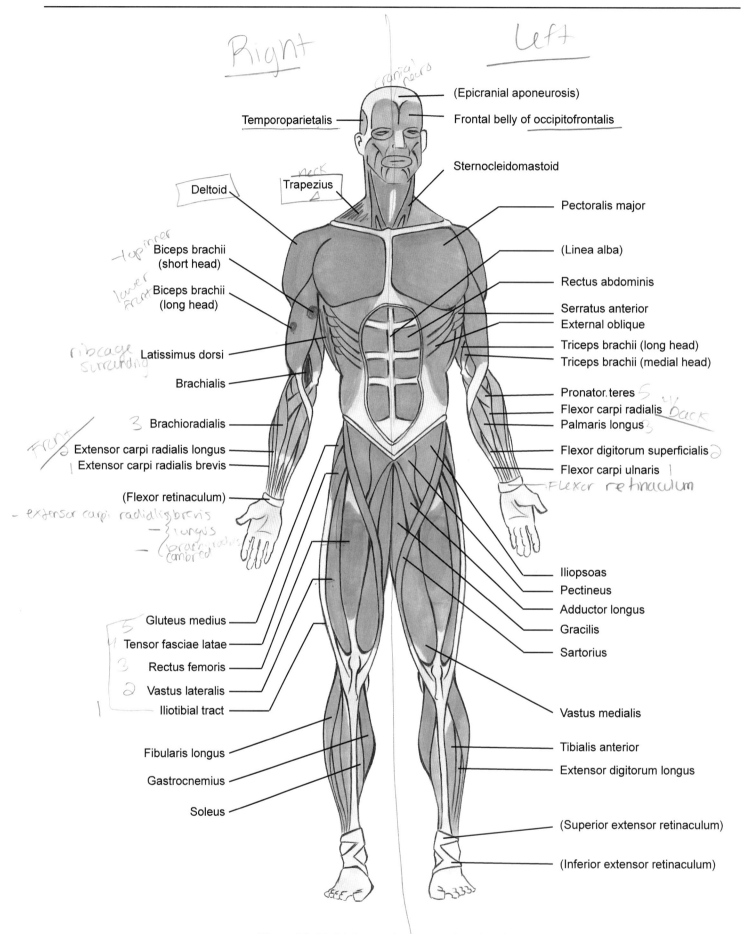

Figure 38. Skeletal muscular system. Anterior view.

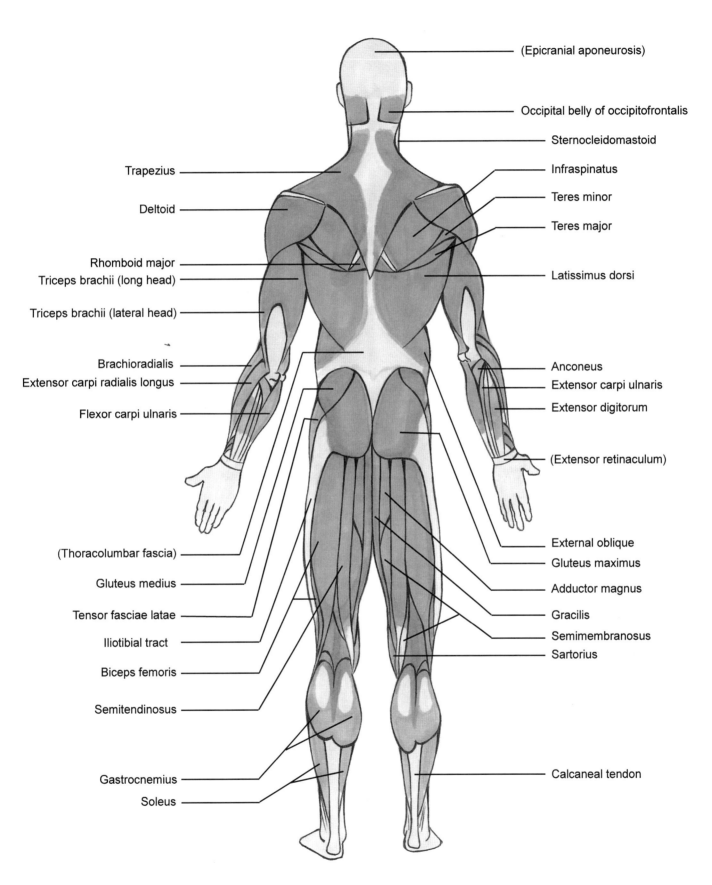

(Epicranial aponeurosis)

Occipital belly of occipitofrontalis

Sternocleidomastoid

Trapezius

Infraspinatus

Teres minor

Deltoid

Teres major

Rhomboid major
Triceps brachii (long head)

Latissimus dorsi

Triceps brachii (lateral head)

Brachioradialis
Extensor carpi radialis longus

Anconeus

Extensor carpi ulnaris

Flexor carpi ulnaris

Extensor digitorum

(Extensor retinaculum)

External oblique

(Thoracolumbar fascia)

Gluteus maximus

Gluteus medius

Adductor magnus

Tensor fasciae latae

Gracilis

Iliotibial tract

Semimembranosus

Biceps femoris

Sartorius

Semitendinosus

Gastrocnemius

Calcaneal tendon

Soleus

Figure 39. Skeletal muscular system. Posterior view.

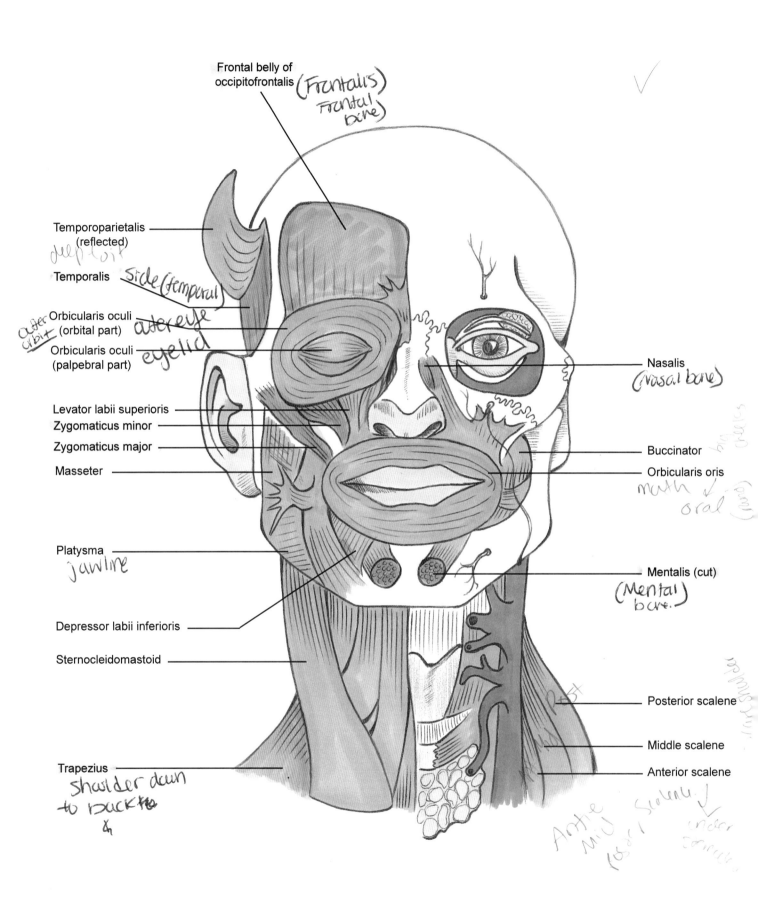

Frontal belly of
occipitofrontalis (Frontalis)
Frontal
bone)

Temporoparietalis
(reflected)
deep bork

Temporalis — Side (temporal)

Orbicularis oculi
(orbital part) above eye
Outer orbit

Orbicularis oculi
(palpebral part) eyelid

Levator labii superioris

Zygomaticus minor

Zygomaticus major

Masseter

Platysma
jawline

Depressor labii inferioris

Sternocleidomastoid

Trapezius
shoulder down
to backto
4

Nasalis
(nasal bone)

Buccinator
big cheeks

Orbicularis oris
mouth
oral (area)

Mentalis (cut)
(Mental)
bone.

Posterior scalene

Middle scalene

Anterior scalene
Ante
Mid
Post Scalene
under corners

Figure 40. Muscles of the head and neck. Anterior view.

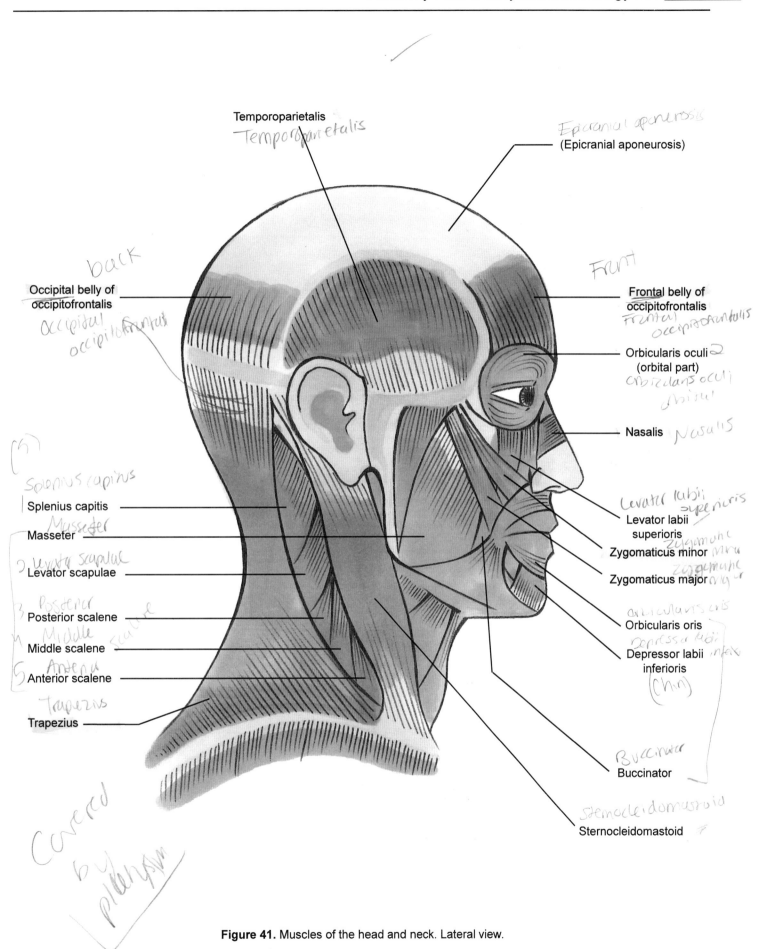

Temporoparietalis

Temporoparietalis

(Epicranial aponeurosis)

Epicranial aponeurosis

back

FRONT

Occipital belly of occipitofrontalis

occipital occipitofrontalis

Frontal belly of occipitofrontalis

Frontal occipitofrontalis

Orbicularis oculi (orbital part)

Orbicularis oculi orbital

Nasalis   Nasalis

(5)

Splenius capitus

Splenius capitis

Masseter

Masseter

2 Levator scapulae

Levator scapulae

Posterior

3 Posterior scalene

Middle scalene

4 Middle scalene

5 Anterior

Anterior scalene

scalene

Trapezius

Trapezius

Covered by platysma

Levator labii superioris

Levator labii superioris

Zygomaticus minor   Zygomatic minor

Zygomaticus major   Zygomatic major

Orbicularis oris

Orbicularis oris

Depressor labii

Depressor labii inferioris (Chin)

Buccinator

Buccinator

Sternocleidomastoid

Sternocleidomastoid

Figure 41. Muscles of the head and neck. Lateral view.

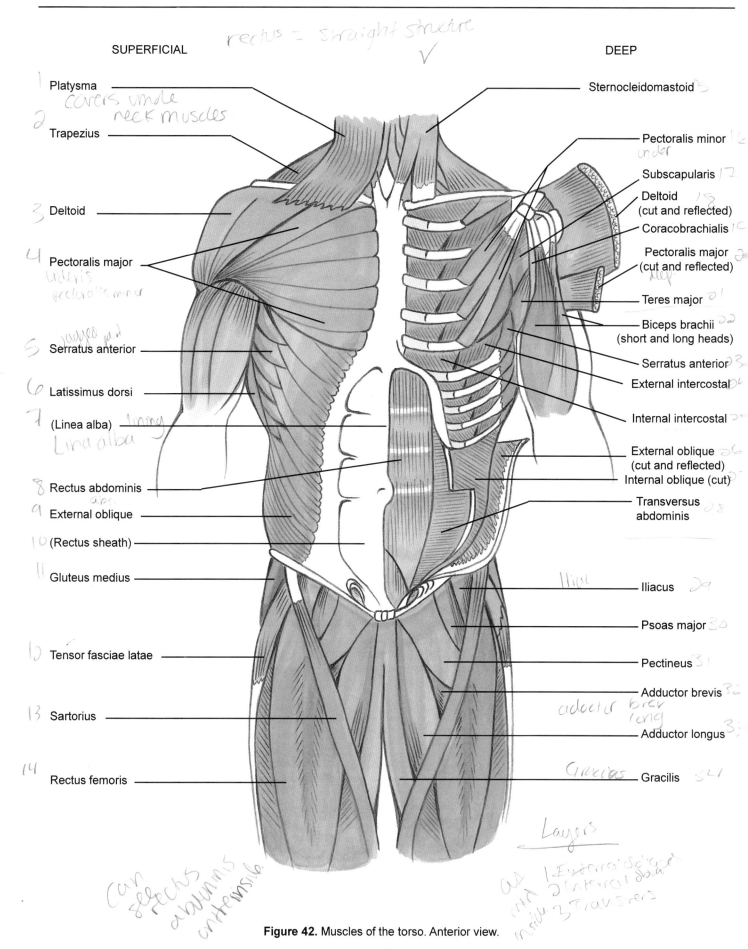

SUPERFICIAL

DEEP

rectus = straight structure

1 Platysma — covers whole neck muscles

2 Trapezius

3 Deltoid

4 Pectoralis major — unders pectoralis minor

5 Serratus anterior — jagged part

6 Latissimus dorsi

7 (Linea alba) — lining — Lina alba

8 Rectus abdominis — abs.

9 External oblique

10 (Rectus sheath)

11 Gluteus medius

12 Tensor fasciae latae

13 Sartorius

14 Rectus femoris

Sternocleidomastoid

Pectoralis minor — under

Subscapularis

Deltoid (cut and reflected)

Coracobrachialis

Pectoralis major (cut and reflected)

Teres major

Biceps brachii (short and long heads)

Serratus anterior

External intercostal

Internal intercostal

External oblique (cut and reflected)

Internal oblique (cut)

Transversus abdominis

Iliacus

Psoas major

Pectineus

Adductor brevis — adductor brev long

Adductor longus

Gracilis

Can effects rectus abdominis anterinsic.

Layers
Q. 1 External oblique
and 2 Internal oblique
inside 3 Transvers

Figure 42. Muscles of the torso. Anterior view.

Front Body

Back ✓

SUPERFICIAL

DEEP

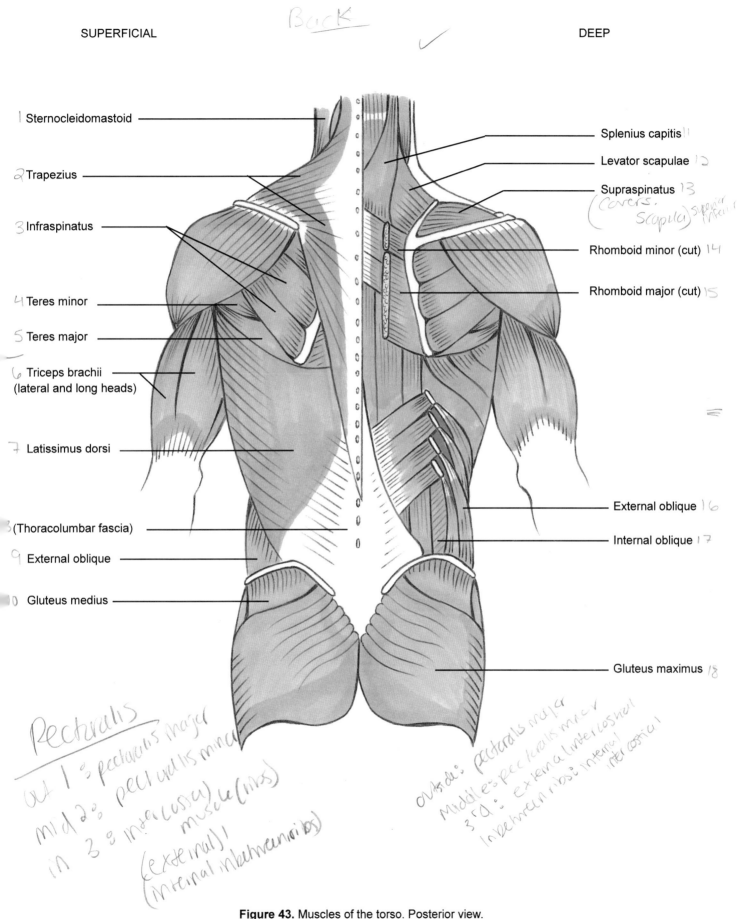

1 Sternocleidomastoid

2 Trapezius

3 Infraspinatus

4 Teres minor

5 Teres major

6 Triceps brachii (lateral and long heads)

7 Latissimus dorsi

8 (Thoracolumbar fascia)

9 External oblique

10 Gluteus medius

Splenius capitis 11

Levator scapulae 12

Supraspinatus 13 (Covers Scapula) superior, inferior.

Rhomboid minor (cut) 14

Rhomboid major (cut) 15

External oblique 16

Internal oblique 17

Gluteus maximus 18

Pectoralis
Out 1° pectoralis major
mid 2° pectoralis minor
in 3° intercostal muscle (ribs)
(external)
(Internal intercostal ribs)

Outside: pectoralis major
Middle: pectoralis minor
3rd: external intercostal
Inbetween rubs: internal intercostal

Figure 43. Muscles of the torso. Posterior view.

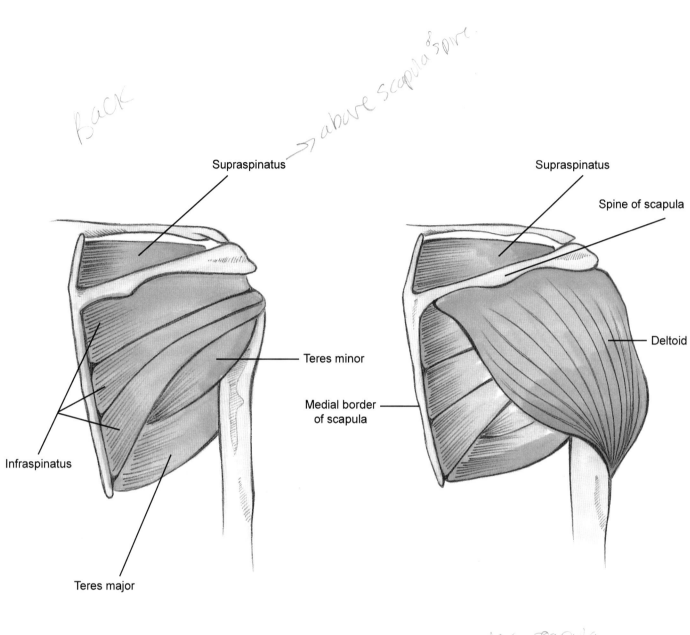

Back

above scapula spine.

Supraspinatus

Supraspinatus

Spine of scapula

Teres minor

Deltoid

Medial border of scapula

Infraspinatus

Teres major

Subscapularis = under scapula.

Figure 44. Muscles of the shoulder girdle. Posterior view.

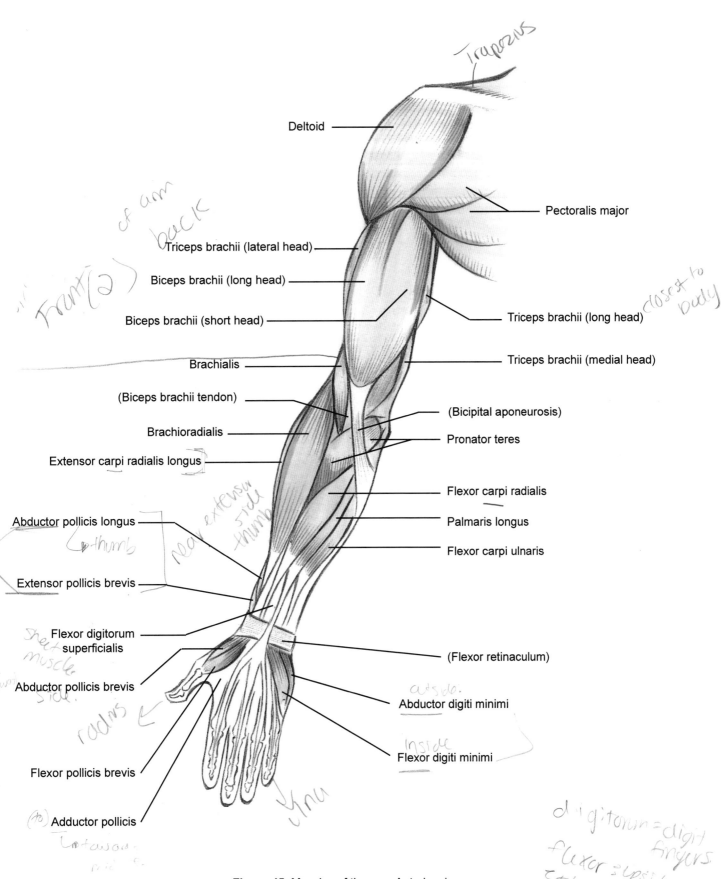

Deltoid

Pectoralis major

Triceps brachii (lateral head)

Biceps brachii (long head)

Biceps brachii (short head)

Triceps brachii (long head)

Brachialis

Triceps brachii (medial head)

(Biceps brachii tendon)

(Bicipital aponeurosis)

Brachioradialis

Pronator teres

Extensor carpi radialis longus

Flexor carpi radialis

Palmaris longus

Abductor pollicis longus

Flexor carpi ulnaris

Extensor pollicis brevis

Flexor digitorum
superficialis

(Flexor retinaculum)

Abductor pollicis brevis

Abductor digiti minimi

Flexor digiti minimi

Flexor pollicis brevis

Adductor pollicis

Figure 45. Muscles of the arm. Anterior view.

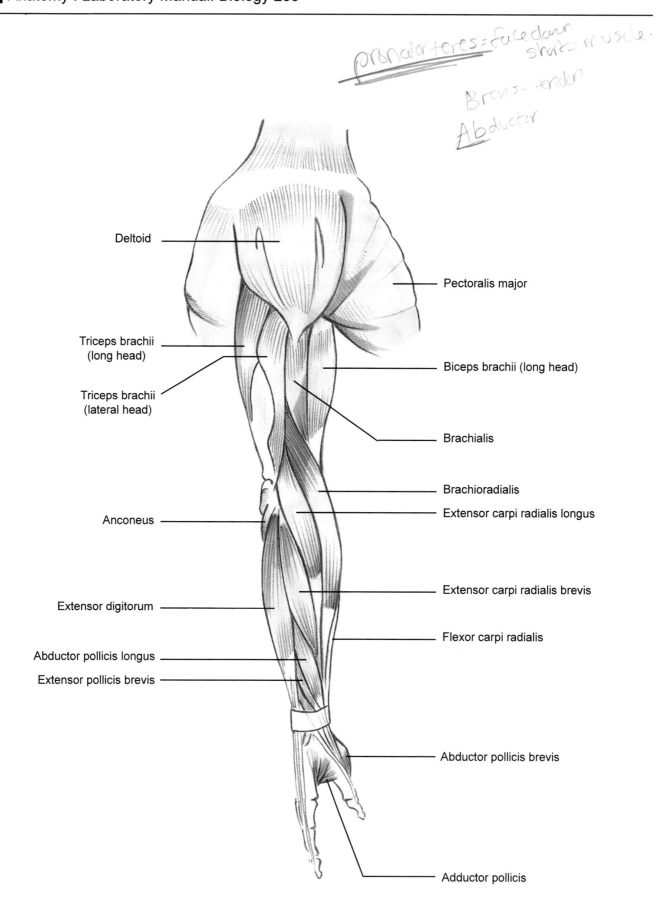

Deltoid

Pectoralis major

Triceps brachii
(long head)

Biceps brachii (long head)

Triceps brachii
(lateral head)

Brachialis

Brachioradialis

Extensor carpi radialis longus

Anconeus

Extensor carpi radialis brevis

Extensor digitorum

Flexor carpi radialis

Abductor pollicis longus

Extensor pollicis brevis

Abductor pollicis brevis

Adductor pollicis

Figure 46. Muscles of the arm. Lateral view.

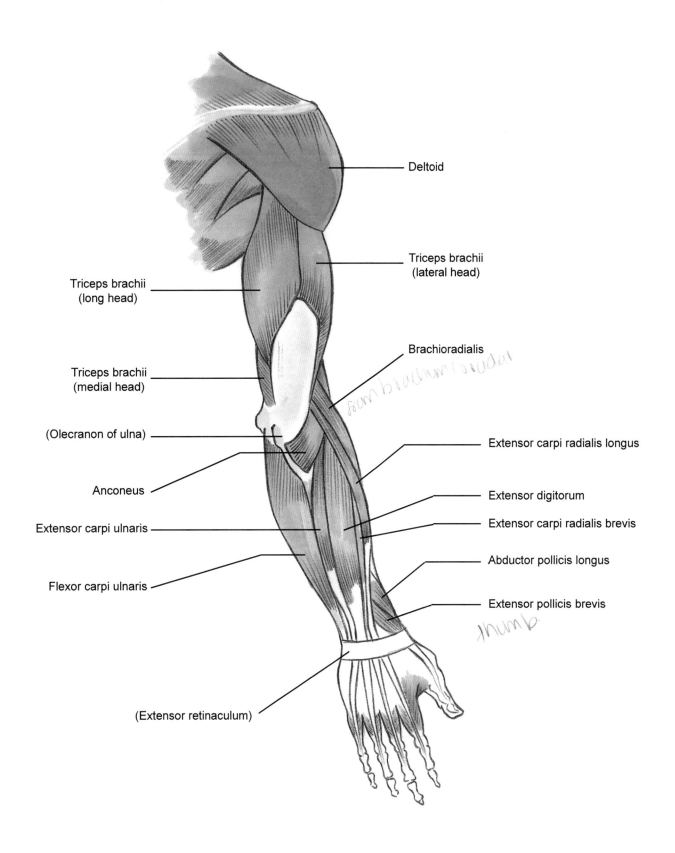

Deltoid

Triceps brachii
(lateral head)

Triceps brachii
(long head)

Brachioradialis

brachium brudul

Triceps brachii
(medial head)

(Olecranon of ulna)

Extensor carpi radialis longus

Anconeus

Extensor digitorum

Extensor carpi ulnaris

Extensor carpi radialis brevis

Abductor pollicis longus

Flexor carpi ulnaris

Extensor pollicis brevis

thumb

(Extensor retinaculum)

Figure 47. Muscles of the arm. Posterior view.

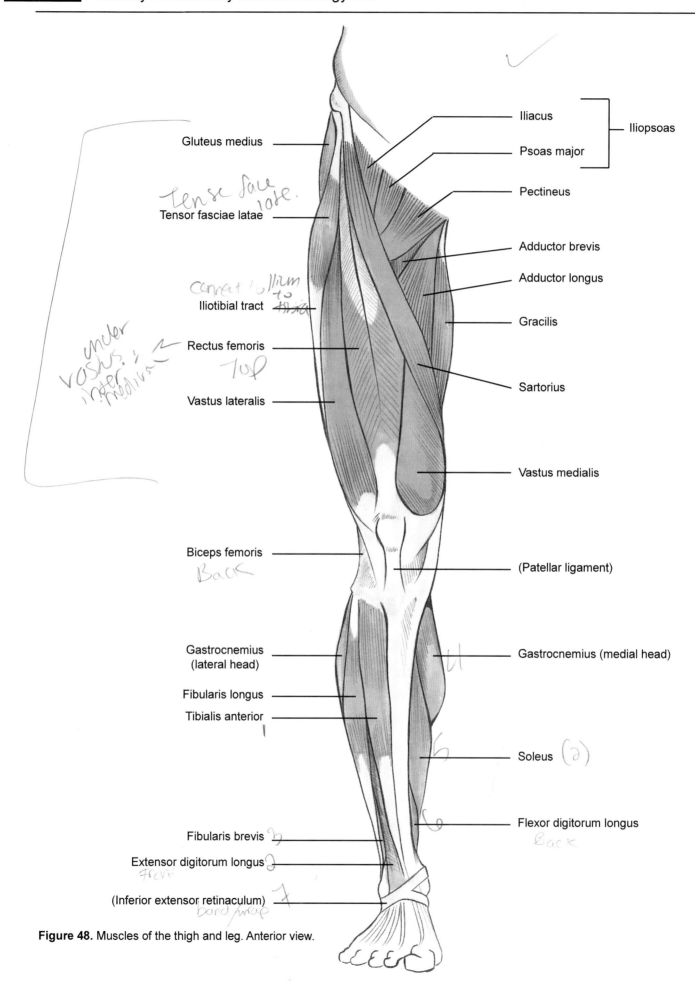

Gluteus medius

Tensor fasciae latae

Tense fasc. late.

correct illium to tibia.

Iliotibial tract

under vastus & inter medius

Rectus femoris

Top

Vastus lateralis

Biceps femoris

Back

Gastrocnemius (lateral head)

Fibularis longus

Tibialis anterior

Fibularis brevis

Extensor digitorum longus

Front

(Inferior extensor retinaculum)

band/wrap

Iliacus

Psoas major

Iliopsoas

Pectineus

Adductor brevis

Adductor longus

Gracilis

Sartorius

Vastus medialis

(Patellar ligament)

Gastrocnemius (medial head)

Soleus

Flexor digitorum longus

Back

Figure 48. Muscles of the thigh and leg. Anterior view.

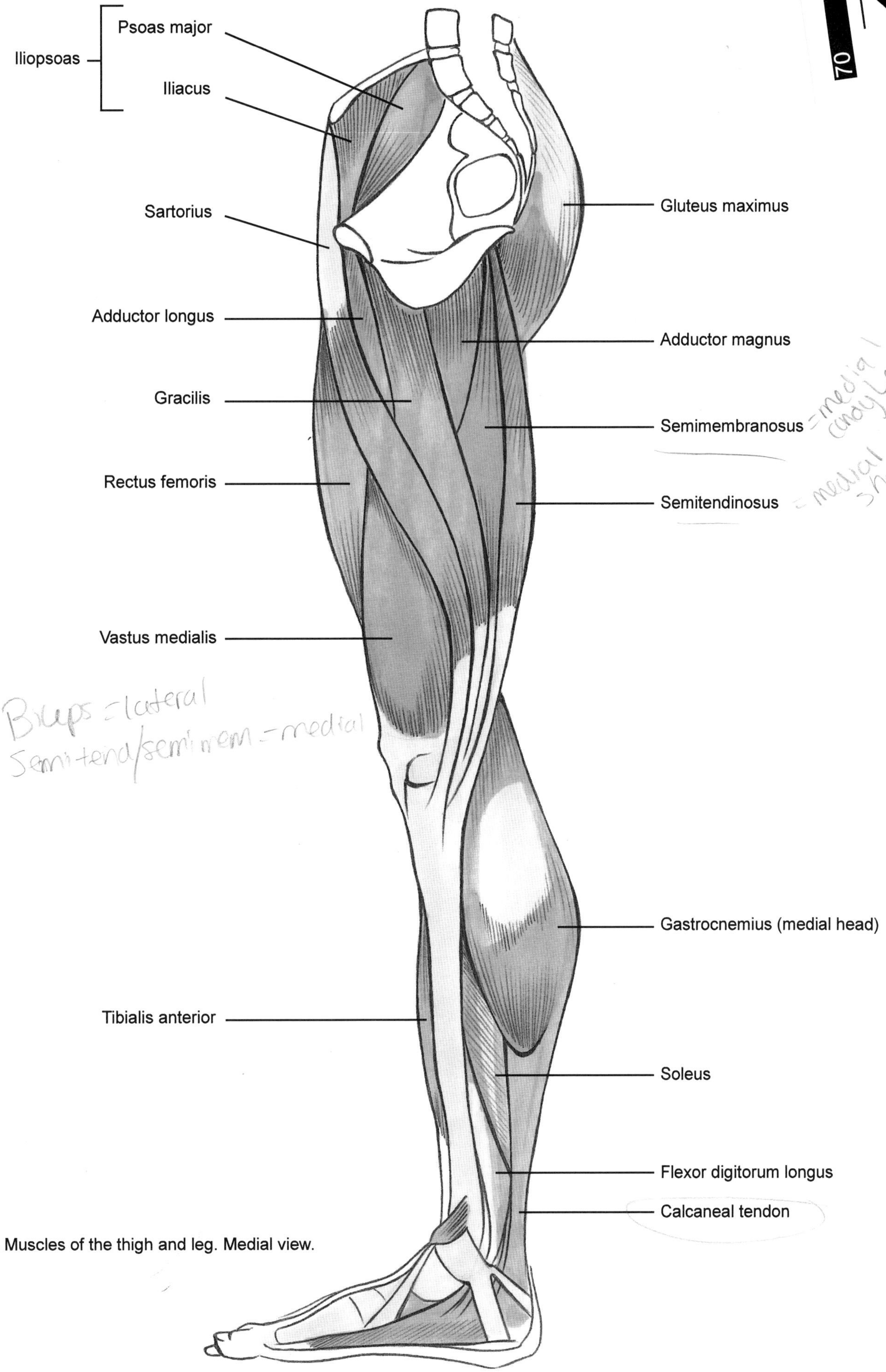

Figure 49. Muscles of the thigh and leg. Medial view.

Iliopsoas
- Psoas major
- Iliacus

Sartorius

Adductor longus

Gracilis

Rectus femoris

Vastus medialis

Gluteus maximus

Adductor magnus

Semimembranosus = medial condyle = medial shaft

Semitendinosus

Tibialis anterior

Gastrocnemius (medial head)

Soleus

Flexor digitorum longus

Calcaneal tendon

* Biceps = lateral
* Semitend/semimem = medial

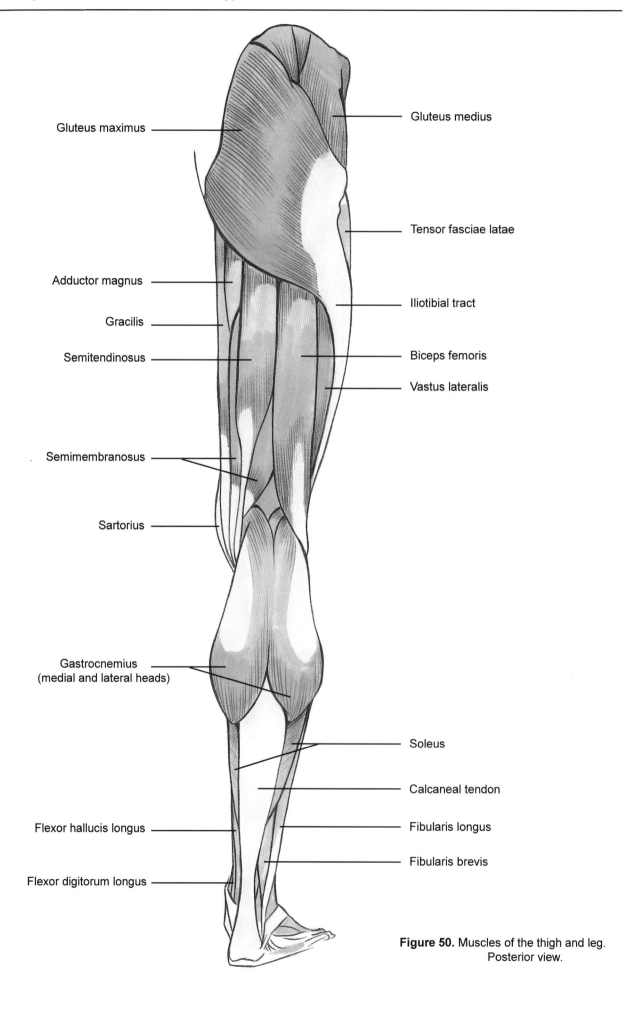

Gluteus maximus

Gluteus medius

Tensor fasciae latae

Adductor magnus

Iliotibial tract

Gracilis

Semitendinosus

Biceps femoris

Vastus lateralis

Semimembranosus

Sartorius

Gastrocnemius
(medial and lateral heads)

Soleus

Calcaneal tendon

Flexor hallucis longus

Fibularis longus

Fibularis brevis

Flexor digitorum longus

Figure 50. Muscles of the thigh and leg.
Posterior view.

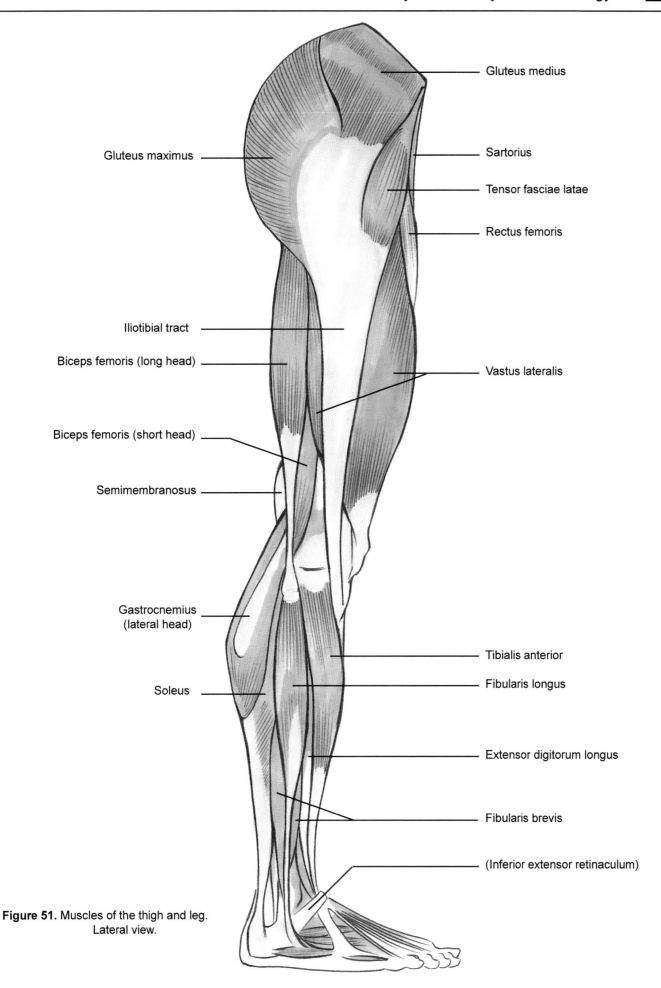

Figure 51. Muscles of the thigh and leg.
Lateral view.

Gluteus medius

Sartorius

Tensor fasciae latae

Rectus femoris

Gluteus maximus

Iliotibial tract

Biceps femoris (long head)

Vastus lateralis

Biceps femoris (short head)

Semimembranosus

Gastrocnemius
(lateral head)

Tibialis anterior

Fibularis longus

Soleus

Extensor digitorum longus

Fibularis brevis

(Inferior extensor retinaculum)

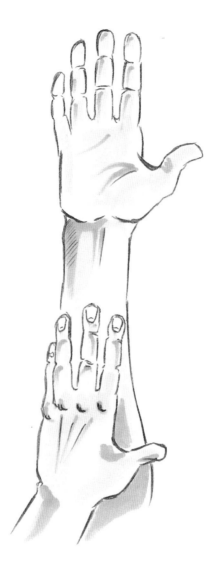

Figure 52. Helpful hints for learning the forearm muscles.

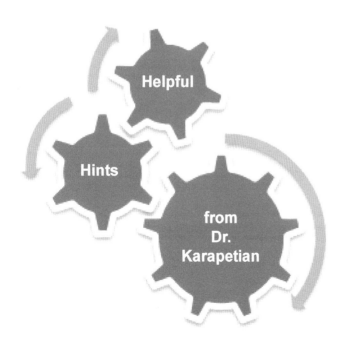

Helpful

Hints

from Dr. Karapetian

To remember some of the muscles of the forearm, put your right arm in front of you, palm up (this is the supine position). Now take your left hand, and set your palm (face down) over the area of your right forearm, close to your elbow (Figure 52) and label your extended (left) fingers as follows:

Digit 1, left thumb: pronator teres
Digit 2, index finger: flexor carpi radialis
Digit 3, middle finger: palmaris longus
Digit 4, ring finger: flexor carpi ulnaris
Digit 5, pinkie finger: extensor carpi ulnaris

Origin, Insertion & Action

The **origin** is the place a muscle attaches, and is usually located on the more proximal bone or stationary point of the body. The **insertion** is the more moveable attachment of a muscle, and is usually located at a more distal point. The insertion is opposite of the origin.

The following pages are origin, insertion, and action charts of the relevant muscles we study in this course; see Tables 12 – 19. I have included the Latin/Greek derivation of the words listed to help make it easier to learn the names of the muscles.

Table 12. Muscles of the Head

MUSCLE	ORIGIN	INSERTION	ACTION
Frontalis (*front* = forehead)	Epicranial aponeurosis	Skin of eyebrow and bridge of nose	Raises eyebrows; wrinkles forehead
Occipitalis (*occipito* = base of skull)	Occipital bone and mastoid process of temporal bone	Epicranial aponeurosis	Tenses and retracts scalp
Orbicularis oculi (*orb* = circular; *oculus* = eye)	Medial margin of orbit	Skin around eyelids	Closes eye
Orbicularis oris (*orb* = circular; *or* = mouth)	Maxilla and mandible	Lips	Compresses, purses lips
Masseter (*maseter* = chewer)	Maxilla and zygomatic arch	Angle and ramus of mandible	Elevates mandible, as in closing mouth, and retracts (draws back) mandible
Temporalis (*tempora* = temples)	Temporal and frontal bones	Coronoid process and ramus of mandible	Elevates and retracts mandible
Levator labii superioris (*levator* = raises; *labii* = lips)	Inferior margin of orbit, superior to the infraorbital foramen	Orbicularis oris	Elevates upper lip
Nasalis (*nasus* = nose)	Maxilla and alar cartilage of nose	Bridge of nose	Compresses bridge, depresses tip of nose; elevates corners of nostrils
Zygomaticus major (*zygomatic* = cheek; *major* = greater)	Zygomatic bone near zygomatico-maxillary suture	Angle of mouth	Retracts and elevates corner of mouth
Zygomaticus minor (*zygomatic* = cheek; *minor* = smaller)	Zygomatic bone posterior to zygomaticotemporal suture	Upper lip	Retracts and elevates upper lip
Risorius (*risor* = laughter)	Fascia over parotid (salivary) gland	Skin at angle of mouth	Draws angle of mouth laterally as in tenseness
Depressor anguli oris (*depressor* = depresses or lowers; *angul* = angle or corner)	Mandible	Angle of mouth	Depresses angle of mouth
Mentalis (*mentum* = chin)	Mandible	Skin of chin	Elevates and protrudes lower lip and pulls skin of chin up as in pouting

Table 13. Muscles of the Neck

MUSCLE	ORIGIN	INSERTION	ACTION
Sternocleidomastoid (*sternum* = breastbone; *cleido* = clavicle; *mastoid* = mastoid process of temporal bone)	Sternum and clavicle	Mastoid process of temporal bone	Acting together (bilateral), flex cervical portion of vertebral column and head; acting singly (unilateral), laterally flex and rotate head to side opposite contracting muscle
Trapezius (*trapezoides* = trapezoid shaped)	Occipital bone, ligamentum nuchae, and spinous processes of thoracic vertebrae	Clavicle and scapula (acromion and scapular spine)	Depends on active region and state of other muscles; may (1) elevate, retract, depress, or rotate scapula upward, (2) elevate clavicle, or (3) extend neck
Splenius capitis (*splenion* = bandage; *caput* = head)	Spinous processes and ligaments connecting seventh cervical and first three (or four) thoracic vertebrae	Occipital bone and mastoid process of temporal bone	Acting together, extend head; acting singly, laterally flex and rotate head to same side as contracting muscle
Levator scapulae (*levator* = raises; *scapulae* = scapula)	Superior four (or five) cervical vertebrae	Superior vertebral border of scapula	Elevates scapula and rotates it downward
Scalenes (anterior, middle, posterior)	Transverse and costal processes of cervical vertebrae	Superior surfaces of first two ribs	Elevates ribs or flexes neck
Platysma (*platy* = flat, broad)	Fascia over deltoid and pectoralis major muscles	Mandible, muscles around angle of mouth, and skin of lower face	Draws outer part of lower lip inferiorly and posteriorly as in pouting; depresses mandible

Table 14. Muscles of the Torso

MUSCLE	ORIGIN	INSERTION	ACTION
Latissimus dorsi (*latissimus* = widest; *dorsum* = back)	Spines of inferior six thoracic vertebrae, lumbar vertebrae, crests of sacrum and ilium, inferior four ribs, inferior angle of scapula	Intertubercular groove of humerus	Extends, adducts, and medially rotates arm at shoulder joint; draws arm inferiorly and posteriorly
Pectoralis major (*pectus* = breast, chest, thorax; *major* = greater)	Clavicle (clavicular head), sternum, and cartilages of second to sixth ribs (sternocostal head)	Greater tubercle and intertubercular groove of humerus	As a whole, adducts and medially rotates arm at shoulder joint; clavicular head alone flexes arm at shoulder joint
Pectoralis minor (*pectus* = breast, chest, thorax; *minor* = lesser)	Third through fifth ribs	Coracoid process of scapula	Depresses and abducts scapula and rotates it downward; elevates third through fifth ribs during forced inspiration when scapula is fixed
Serratus anterior (*serratus* = saw toothed; *anterior* = front)	Superior eight (or nine) ribs	Vertebral border and inferior angle of scapula	Abducts scapula and rotates it upward; elevates ribs when scapula is fixed
External oblique (*external* = closer to surface; *oblique* = neither parallel nor perpendicular)	Inferior eight ribs	Iliac crest and linea alba	Acting together (bilateral), compress abdomen and flex vertebral column; acting singly (unilateral), laterally flex vertebral column, especially lumbar portion, and rotate vertebral column
Internal oblique (*internal* = farther from surface; *oblique* = neither parallel nor perpendicular)	Iliac crest, inguinal ligament, and thoracolumbar fascia	Cartilage of last three (or four) ribs and linea alba	Acting together, compress abdomen and flex vertebral column; acting singly, laterally flex vertebral column, especially lumbar portion, and rotate vertebral column
Rectus abdominis (*rectus* = straight, fibers parallel to midline; *abdomino* = abdomen)	Pubic crest and pubic symphysis	Cartilage of fifth to seventh ribs and xiphoid process	Flexes vertebral column, especially lumbar portion, and compresses abdomen to aid in defecation, urination, forced expiration, and parturition (childbirth)
Transversus abdominis (*transverse* = perpendicular to midline)	Iliac crest, inguinal ligament, lumbar fascia, and cartilages of inferior six ribs	Xiphoid process, linea alba, and pubis	Compresses abdomen
Rhomboid major (*rhomboides* = rhomboid or diamond shaped; *major* = greater)	Spines of second to fifth thoracic vertebrae	Vertebral border of scapula inferior to spine	Elevates and adducts scapula and rotates it downward; stabilizes scapula
Rhomboid minor (*rhomboides* = rhomboid or diamond shaped; *minor* = smaller)	Spines of seventh cervical and first thoracic vertebrae	Vertebral border of scapula superior to spine	Elevates and adducts scapula and rotates it downward; stabilizes scapula
Internal intercostals (*internal* = farther from surface; *costa* = rib)	Superior border of rib below	Inferior border of rib above	Draw adjacent ribs together during forced expiration and thus decrease lateral and anteroposterior dimensions of thorax
Diaphragm (*dia* = across; *phragma* = wall)	Xiphoid process of the sternum, costal cartilages of inferior six ribs, and lumbar vertebrae	Central tendon	Forms floor of thoracic cavity; pulls central tendon inferiorly during inspiration, and as dome of diaphragm flattens increases vertical length of thorax

Table 15. Muscles of the Shoulder Girdle

MUSCLE	ORIGIN	INSERTION	ACTION
Supraspinatus (*supra* = above; *spinatus* = spine of scapula)	Supraspinous fossa of scapula	Greater tubercle of humerus	Assists deltoid muscle in abducting arm at shoulder joint
Infraspinatus (*infra* = bleow; *spinatus* = spine of scapula)	Infraspinous fossa of scapula	Greater tubercle of humerus	Laterally rotates and adducts arm at shoulder joint
Teres major (*teres* = long and round; *major* = greater)	Inferior angle of scapula	Intertubercular groove of humerus	Extends arm at shoulder joint and assists in adduction and medial rotation of arm at shoulder joint
Teres minor (*teres* = long and round; *minor* = smaller)	Inferior lateral border of scapula	Greater tubercle of humerus	Laterally rotates, extends, and adducts arm at shoulder joint
Subscapularis (*sub* = below; *scapularis* = scapula)	Subscapular fossa of scapula	Lesser tubercle of humerus	Medially rotates arm at shoulder joint
Deltoid (*delta* = triangular)	(1) Anterior fibers: acromial extremity of clavicle, (2) Middle fibers: acromion of scapula, (3) Posterior fibers: spine of scapula	Deltoid tuberosity of humerus	Middle fibers abduct arm at shoulder joint; anterior fibers flex and medially rotate arm at shoulder joint; posterior fibers extend and laterally rotate arm at shoulder joint

Table 16. Muscles of the Arm

MUSCLE	ORIGIN	INSERTION	ACTION
Biceps brachii (*biceps* = two heads of origin; *brachion* = arm)	(1) Long head: supraglenoid tubercle of the scapula, (2) Short head: coracoid process of scapula	Radial tuberosity and bicipital aponeurosis	Flexes forearm at elbow joint, supinates forearm at radioulnar joints, and flexes arm at shoulder joint
Brachialis (*brachion* = arm)	Distal, anterior surface of humerus	Ulnar tuberosity and coronoid process of ulna	Flexes forearm at elbow joint
Triceps brachii (*triceps* = three heads of origin; *brachion* = arm)	(1) Long head: projection inferior to glenoid cavity of scapula, (2) Lateral head: lateral and posterior surface of humerus superior to radial groove, (3) Medial head: entire posterior surface of humerus inferior to a groove for the radial nerve	Olecranon of ulna	Extends forearm at elbow joint and extends arm at shoulder joint
Coracobrachialis (*coraco* = coracoid process)	Coracoid process of scapula	Middle of medial surface of shaft of humerus	Flexes and adducts arm at shoulder joint

Table 17. Muscles of the Forearm

MUSCLE	ORIGIN	INSERTION	ACTION
Pronator teres (*pronation* = turning palm downward or posteriorly)	Medial epicondyle of humerus and coronoid process of ulna	Midlateral surface of radius	Pronates forearm at radioulnar joint and weakly flexes forearm at elbow joint
Flexor carpi radialis (*flexor* = decreases angle at joint; *carpus* = wrist; *radialis* = radius)	Medial epicondyle of humerus	Second and third metacarpals	Flexes and abducts hand at wrist joint
Palmaris longus (*palma* = palm; *longus* = long)	Medial epicondyle of humerus	Flexor retinaculum and palmar aponeurosis (deep fascia in center of palm)	Weakly flexes hand at wrist joint
Flexor carpi ulnaris (*flexor* = decreases angle at joint; *carpus* = wrist; *ulnaris* = ulna)	Medial epicondyle of humerus and superior posterior border of ulna	Pisiform, hamate, and fifth metacarpal	Flexes and adducts hand at wrist joint
Extensor carpi ulnaris (*extensor* = increases angle at joint; *carpus* = wrist; *ulnaris* = ulna)	Lateral epicondyle of humerus and posterior border of ulna	Fifth metacarpal	Extends and adducts hand at wrist joint
Extensor digitorum (*extensor* = increases angle at joint; *digit* = finger or toe)	Lateral epicondyle of humerus	Distal and middle phalanges of each finger	Extends distal and middle phalanges of each finger at interphalangeal joints, proximal phalanx of each finger at metacarpophalangeal joint, and hand at wrist
Extensor carpi radialis brevis (*extensor* = increases angle at joint; *carpus* = wrist; *radialis* = radius; *brevis* = short)	Lateral epicondyle of humerus	Third metacarpal	Extends and abducts hand at wrist joint
Extensor carpi radialis longus (*extensor* = increases angle at joint; *carpus* = wrist; *radialis* = radius; *longus* = long)	Lateral supracondylar ridge of humerus	Second metacarpal	Extends and abducts hand at wrist joint
Brachioradialis (*brachion* = arm; *radialis* = radius)	Medial and lateral borders of distal end of humerus	Superior to styloid process of radius	Flexes forearm at elbow joint and supinates and pronates forearm at radioulnar joints to neutral position
Extensor pollicis brevis (*extensor* = increases angle at joint; *pollex* = thumb; *brevis* = short)	Posterior surface of middle of radius and interosseous membrane	Base of proximal phalanx of thumb	Extends proximal phalanx of thumb at metacarpophalangeal joint, first metacarpal of thumb at carpometacarpal joint, and hand at wrist
Abductor pollicis longus (*abductor* = moves part away from midline; *pollex* = thumb; *longus* = long)	Posterior surface of middle of radius and ulna and interosseous membrane	First metacarpal	Abducts and extends thumb at carpometacarpal joint and abducts hand at wrist
Anconeus (*anconeal* = pertaining to elbow)	Lateral epicondyle of humerus	Olecranon and superior portion of shaft of ulna	Extends forearm at elbow joint
Flexor digitorum superficialis (*flexor* = decreases angle at joint; *digit* = finger or toe)	Medial epicondyle of humerus; adjacent anterior surfaces of ulna and radius	Midlateral surface of middle phalanges of fingers 2-5	Flexion at proximal interphalangeal, metacarpophalangeal, and wrist joints
Supinator (*supination* = turning palm upward or anteriorly)	Lateral epicondyle of humerus and ridge near radial notch of ulna	Lateral surface of proximal one-third of radius	Supinates forearm at radioulnar joints

Table 18. Muscles of the Hip and Thigh

MUSCLE	ORIGIN	INSERTION	ACTION
Psoas major (*psoa* = muscle of loin; *major* = greater)	Transverse processes and bodies of lumbar vertebrae	With iliacus into lesser trochanter of femur	Both psoas major and iliacus muscles acting together flex thigh at hip joint, rotate thigh laterally, and flex trunk on the hip as in sitting up from the supine position
Iliacus (*iliac* = ilium)	Iliac fossa	With psoas major into lesser trochanter of femur	(same as above)
Gluteus maximus (*glutos* = buttock; *maximus* = largest; strongest single muscle in body)	Iliac crest, sacrum, coccyx, and aponeurosis of sacrospinalis	Iliotibial tract of fascia latae and lateral part of linea aspera under greater trochanter (gluteal tuberosity) of femur	Extends thigh at hip joint and laterally rotates thigh
Gluteus medius (*glutos* = buttock; *media* = middle)	Ilium	Greater trochanter of femur	Abducts thigh at hip joint and medially rotates thigh
Tensor fasciae latae (*tensor* = makes tense; *fascia* = band; *latus* = wide)	Iliac crest	Tibia by way of the iliotibial tract	Flexes and abducts thigh at hip joint
Sartorius (*sartor* = tailor; longest muscle in the body)	Anterior superior iliac spine	Medial surface of body of tibia	Flexes leg at knee joint; flexes, abducts, and laterally rotates thigh at hip joint
Rectus femoris (*rectus* = fibers parallel to midline; *femoris* = femur)	Anterior inferior iliac spine	Patella via quadriceps tendon and then tibial tuberosity via patellar ligament	All four heads (of the quadriceps) extend leg at knee joint; rectus femoris muscle acting alone also flexes thigh at hip joint
Vastus medialis (*vastus* = large; *medialis* = medial)	Linea aspera of femur	(same as above)	(same as above)
Vastus lateralis (*vastus* = large; *lateralis* = lateral)	Greater trochanter and linea aspera of femur	(same as above)	(same as above)
Vastus intermedius (*vastus* = large; *intermedius* = middle)	Anterior and lateral surfaces of body of femur	(same as above)	(same as above)
Pectineus (*pecten* = comb-shaped)	Superior ramus of pubis	Pectineal line of femur, between lesser trochanter and linea aspera	Flexes and adducts thigh at hip joint
Adductor longus (*adductor* = moves part closer to midline; *longus* = long)	Pubic crest and pubic symphysis	Linea aspera of femur	Adducts and flexes thigh at hip joint and medially rotates thigh
Adductor magnus (*adductor* = moves part closer to midline; *magnus* = large)	Inferior ramus of pubis and ischium to ischial tuberosity	Linea aspera of femur	Adducts thigh at hip joint and medially rotates thigh; anterior part flexes thigh at hip joint, posterior part extends thigh at hip joint
Gracilis (*gracilis* = slender)	Pubic symphysis and pubic arch	Medial surface of body of tibia	Adducts thigh at hip joint, medially rotates thigh, and flexes leg at knee joint
Biceps femoris (*biceps* = two heads of origin; *femoris* = femur)	Long head arises from ischial tuberosity; short head arises from linea aspera of femur	Head of fibula and lateral condyle of tibia	(The hamstrings) flex leg at knee joint and extend thigh at hip joint
Semitendinosus (*semi* = half; *tendo* = tendon)	Ischial tuberosity	Proximal part of medial surface of shaft of tibia	(same as above)
Semimembranosus (*semi* = half; *membran* = membrane)	Ischial tuberosity	Medial condyle of tibia	(same as above)

Table 19. Muscles of the Leg

MUSCLE	ORIGIN	INSERTION	ACTION
Gastrocnemius (*gaster* = belly; *kneme* = knee)	Lateral and medial condyles of femur and capsule of knee	Calcaneus by way of calcaneal (Achilles) tendon	Plantar flexes foot and leg at knee join
Soleus (*soleus* = sole of foot)	Head of fibula and medial border of tibia	Calcaneus by way of calcaneal (Achilles) tendon	Plantar flexes foot at ankle joint
Plantaris (*plantar* = sole of foot)	Femur superior to lateral condyle	Calcaneus by way of calcaneal (Achilles) tendon	Plantar flexes foot at ankle joint and flexes leg at knee joint
Tibialis anterior (*tibialis* = tibia; *anterior* = front)	Lateral condyle and body of tibia and interosseous membrane (sheet of fibrous tissue that holds shafts of tibia and fibula together)	First metatarsal and first (medial) cuneiform	Dorsiflexes foot at ankle joint and inverts foot at intertarsal joints
Extensor digitorum longus (*extensor* = increases angle at joint; *digit* = finger or toe; *longus* = long)	Lateral condyle of tibia, anterior surface of fibula, and interosseous membrane	Middle and distal phalanges of toes 2-5	Dorsiflexes foot at ankle joint, everts foot at intertarsal joints, and extends distal and middle phalanges of each toe at interphalangeal joints and proximal phalanx of each toe at metatarsophalangeal joint
Fibularis (Peroneus) longus (*perone* = fibula; *longus* = long)	Head and body of fibula and lateral condyle of tibia	First metatarsal and first (medial) cuneiform	Plantar flexes foot at ankle joint and everts foot at intertarsal joints
Fibularis (Peroneus) brevis (*perone* = fibula; *brevis* = short)	Body of fibula	Fifth metatarsal	Plantar flexes foot at ankle joint and everts foot at intertarsal joints
Flexor digitorum longus (*flexor* = decreases angle at joint; *digit* = finger or toe; *longus* = long)	Posteromedial surface of tibia	Inferior surfaces of distal phalanges, toes 2-5	Flexion at joints of toes 2-5
Flexor hallucis longus (*flexor* = decreases angle at joint; *hallux* = great toe; *longus* = long)	Posterior surface of fibula	Inferior surface of distal phalanx of great toe	Flexion at joints of great toe

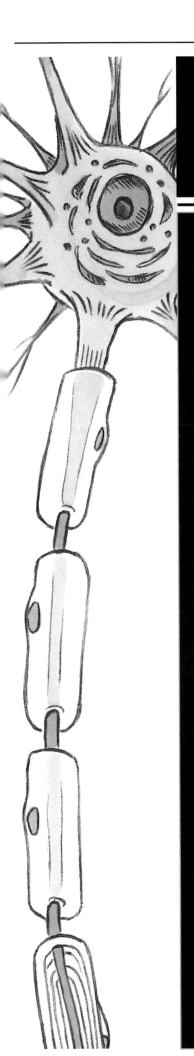

UNIT III

Introduction

he final unit (At last! I thought this class would go on forever!) we will cover is neurology (neuron = nerve; -logy = study of). The two main divisions of the nervous system are the **central nervous system** (**CNS**) and the **peripheral nervous system** (**PNS**). The CNS includes the brain and spinal cord, while the PNS includes the cranial nerves that arise from the brain and the spinal nerves that bud from the spinal cord. The **autonomic nervous system** (**ANS**) is considered a functional subdivision of the CNS, as its organizing centers are located within the brain. The peripheral parts of the ANS are further divided into the **sympathetic** and **parasympathetic divisions**. These divisions usually have opposing effects on the same organs – while the sympathetic response may stimulate (fight or flight), the parasympathetic response typically inhibits (rest and digest).

The cell bodies within the CNS usually gather into areas called nuclei (not to be confused with the nucleus of a cell). Cell bodies in the PNS usually occur in bunches called **ganglia** (*ganglion* = knot on a string, swelling).

The branched extensions from the cell body are called **dendrites** (*dendron* = tree branch), they receive stimuli and conduct impulses to the cell body. Dendritic spinules are the miniature extensions that cover some dendrites in an effort to increase surface area and contact with other neurons. The **axon** (*axon* = axis) is a relatively long, tube-like process that conducts impulses away from the cell body. Axons range from a few millimeters in the CNS to over a meter in length when stretching from the spinal cord to our distal extremities. The cytoplasm of an axon contains microtubules, mitochondria, and neurofibrils.

Neural tissue

Multipolar Neuron

he basic functional and structural unit of the nervous system is the **neuron**, a specialized type of cell that can conduct an impulse, release specific chemicals, and has the ability to respond to both chemical and physical stimuli (Figure 53). Neurons contain three basic components: a cell body, dendrites, and an axon.

The neuronal **cell body** is an enlarged portion of the neuron where the nucleus and nucleolus are surrounded by cytoplasm. But in addition to the typical organelles found in most cells, the cytoplasm of a neuron also contains **Nissl bodies** (Franz Nissl, German 1860-1919) also called **chromatophilic substances** which are specialized layers of rough endoplasmic reticulum which function in protein synthesis, and microtubules which transport materials within the cell.

When a neurological cell containing a white lipid-protein substance called **myelin** (*myelos* = marrow) wraps around portions of the axon, the process is called myelination. Myelinated neurons can be found in both the CNS (by way of oligodendrocytes) and the PNS (by way of neurolemmocytes) where the myelin aides in the conduction of impulses and provides support. Some neurons, however, are unmyelinated.

yelin is the source of **white matter** in the brain and spinal cord and causes the white coloration of nerves. In the PNS, **Schwann cells** (Theodor Schwann, German 1810-1882) or **neurolemmocytes** form myelin layers around the axons located there. Each neurolemmocyte wraps about 1mm of axon creating a **myelinated internode**, and in the process leave gaps called **nodes of Ranvier** (Louis Ranvier, French 1835-1922) or **neurofibril nodes** which separate each of the internodes. **Neurolemma** (*neuron* = nerve; *lemma* = husk) enclose the myelin sheath and may help promote neuron regeneration if an injury occurs – pretty impressive. The telodendria of axons end at synaptic terminals, the location of **synaptic knobs**. This is the site of much neurotransmitter activity, and where neurons communicate with other cells (Figure 54).

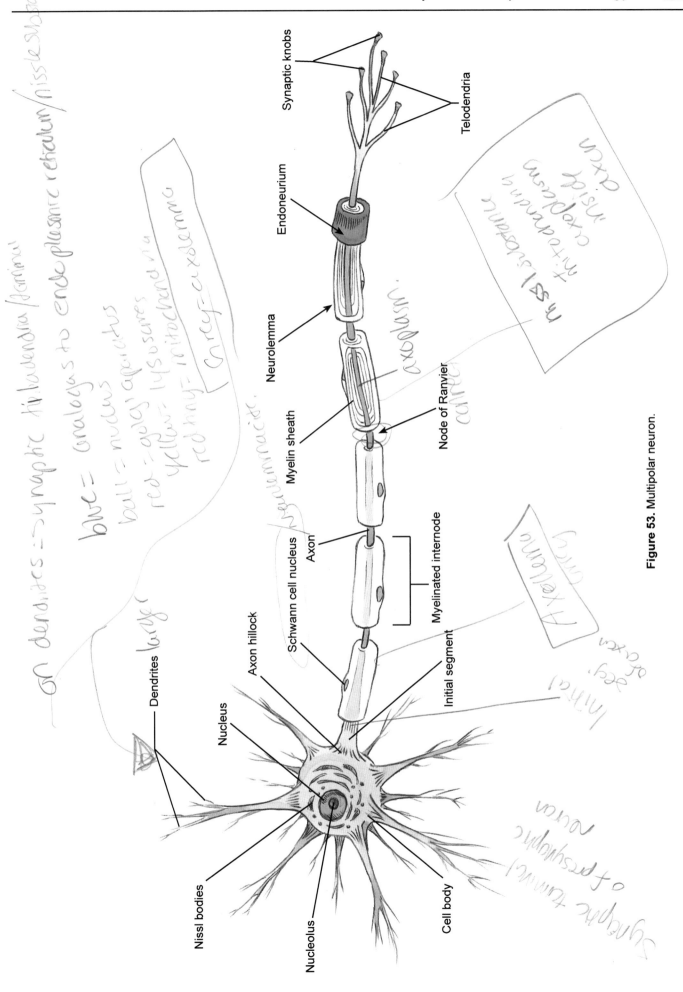

Figure 53. Multipolar neuron.

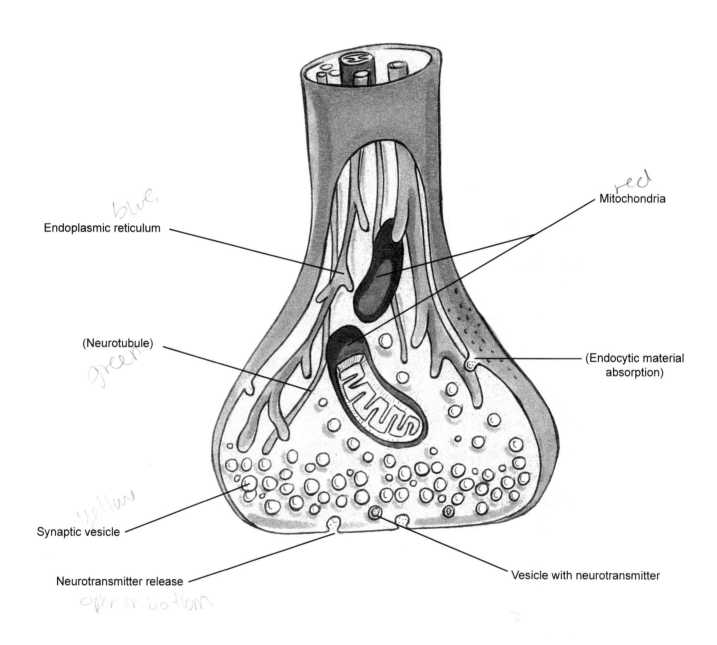

Endoplasmic reticulum

blue

Mitochondria

red

(Neurotubule)

green

(Endocytic material absorption)

Synaptic vesicle

yellow

Neurotransmitter release

open vacuolum

Vesicle with neurotransmitter

Figure 54. Synaptic knob.

Related Structures of the CNS

Meninges

The CNS is a delicate structure, which is why it is encased in bone – the cranium surrounds the brain and the vertebral column surrounds the spinal cord. But in addition there are three connective tissue coverings called the **meninges** (*mening* = membrane). From superficial to deep they are the dura mater, arachnoid mater, and pia mater (Figure 55 and 56).

Dura Mater

The **dura mater** (*dura* = hard; *mater* = mother) is made up of dense fibrous connective tissue and connects to the bone that surrounds it... I always loved this one because I imagine all the body parts having their formal annual meeting and the dura mater always gets to be introduced as the *hard mother!* How awesome is that? The cranial dura mater has a thick outer periosteal (pair-ē-OS-teal) layer that attaches to the skull, and a thin inner meningeal layer that envelops the overall shape of the brain – the cranial dura mater is therefore a double-layer. The spinal dura mater is composed of similar structures, however, this dural sheath is single layered and does not connect to the vertebrae, instead, an **epidural space** forms (Figure 56).

Arachnoid Mater

The **arachnoid mater** (*arachnida* = spider, as in cobweb; *mater* = mother) does not sound nearly as cool as the hard mother. Oooooh cobwebs! Really though, it is the middle layer of the three meninges. It is a netlike membrane that spreads over the CNS, but does not extend into the sulci or fissures of the brain. The **subarachnoid space** is where **cerebrospinal fluid** circulates. Superficial to the brain, knob-like projections called **arachnoid villi** protrude through the dura mater allowing cerebrospinal fluid to be absorbed into the venous blood of the dural sinuses above the brain.

Pia Mater

The **pia mater** (*pia* = soft, delicate, tender; *mater* = mother) just sounds plain wimpy. I mean if this layer got into a fight with the dura mater, it would be over in three seconds. Anyway, it is the deepest layer which follows the contours of the brain and spinal cord. It is quite vascular because it helps to nourish the underlying cells of the brain and spinal cord. It becomes somewhat specialized as it overlays the roofs of the ventricles (I will introduce these soon) where it contributes to the formation of the choroid plexuses (I will discuss these too). Extensions called **denticulate ligaments** protrude from the pia mater to help attach the spinal cord to the dura mater.

Cranial Meninges: Dural Septa of the Cranium

 xtensions of the dura mater reach inward to form flat septa, or walls, that help to anchor the brain in place (Figure 57). I like to refer to these as the seatbelts of the skull – keeping the brain safe within the cranium, in the same way seatbelts keep us safe in our cars. They include the:

- **Falx cerebri** (FALL-ks; *falks* = sickle) – a midsagittal, sickle shaped fold that extends the length of the longitudinal fissure between the two hemispheres. The **superior sagittal sinus** and the **inferior sagittal sinus** lie within this dural fold.

- **Falx cerebelli** – another vertical fold in the sagittal plane that partly divides the cerebellum into two hemispheres.

- **Tentorium cerebelli** (*tentorium* = a covering, tent) – a horizontal fold that separates the cerebrum (on top) from the cerebellum (underneath). The **transverse sinus** lies within the tentorium cerebelli.

Blood supply

A highly organized system of cerebral arteries supplies blood to the brain. Adequate blood flow is essential here because interruptions may lead to the loss of consciousness (after a few seconds) and permanent brain damage (after fewer than five minutes). The left and right **internal carotid arteries** supply the front half of the cerebrum while the **vertebral arteries** supply the back of the cerebrum, the cerebellum, and the brainstem.

The internal carotid arteries and the **basilar artery** are connected with the three paired **cerebral arteries** (anterior, middle, and posterior) and the **communicating arteries** (anterior and posterior) to form the **cerebral arterial circle** (Figure 58) also known as the **circle of Willis** (Thomas Willis, British 1621-1675).

Cerebral Arterial Circle:
- Anterior communicating artery
- Anterior cerebral artery (left & right)
- Middle cerebral artery (left & right)
- Posterior communicating artery (left & right)
- Posterior cerebral artery (left & right)

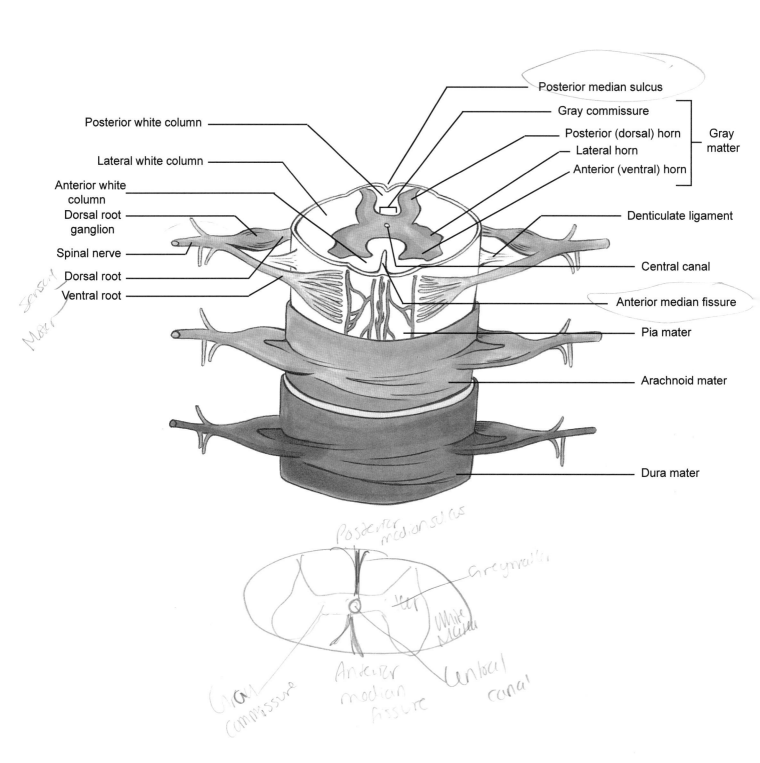

Figure 55. Sectional anatomy of spinal cord. Three-dimensional view of adult spinal cord and meninges.

Yellow = dorsal root ganglia (big one)

Ventral root = under dorsal ganglia no head
Dorsal root = extension of ganglia
red = vertebral arteries

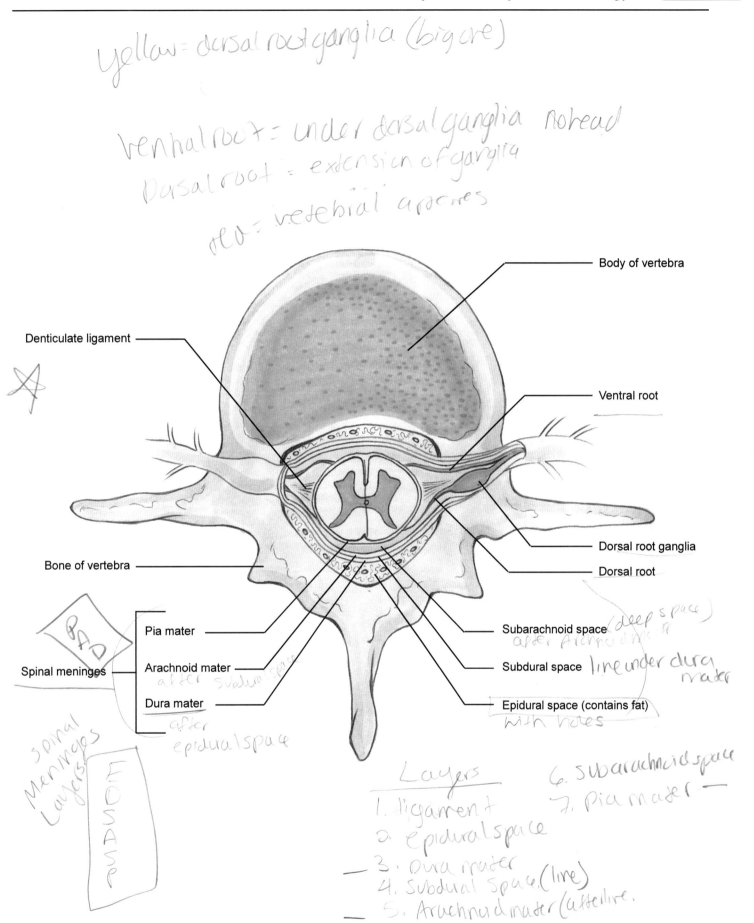

Body of vertebra

Denticulate ligament

Ventral root

Bone of vertebra

Dorsal root ganglia

Dorsal root

Pia mater

Subarachnoid space (deep space) after Arachnoid m.

Spinal meninges

Arachnoid mater after subdural space

Subdural space line under dura mater

Dura mater after epidural space

Epidural space (contains fat) with holes

PAD

Spinal Meninges Layers

EDSASP

Layers
1. ligament
2. epidural space
3. Dura mater
4. Subdural space (line)
5. Arachnoid mater (after line.
6. subarachnoid space
7. Pia mater —

Figure 56. Anatomy of spinal cord. Cross-section of spinal cord and surrounding vertebral column.

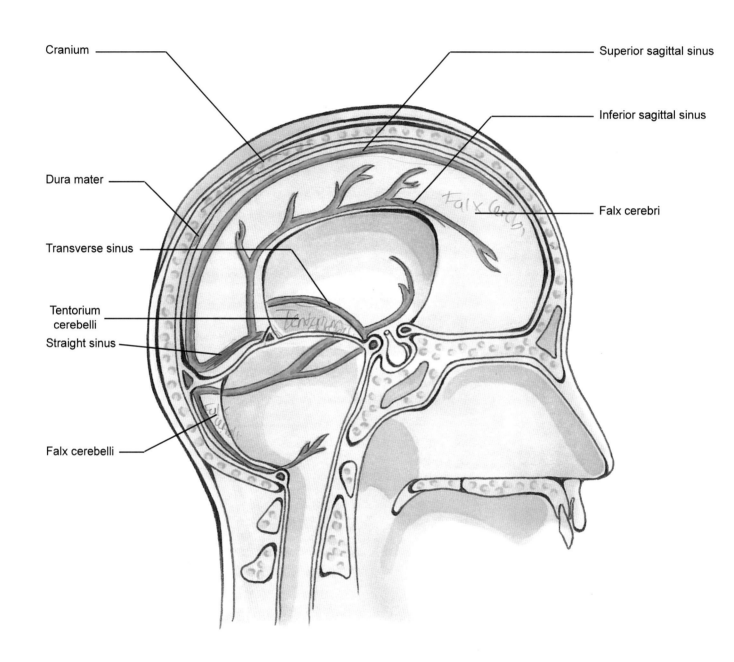

Cranium

Dura mater

Transverse sinus

Tentorium
cerebelli

Straight sinus

Falx cerebelli

Superior sagittal sinus

Inferior sagittal sinus

Falx cerebri

Figure 57. Cranial meninges. Sagittal view.

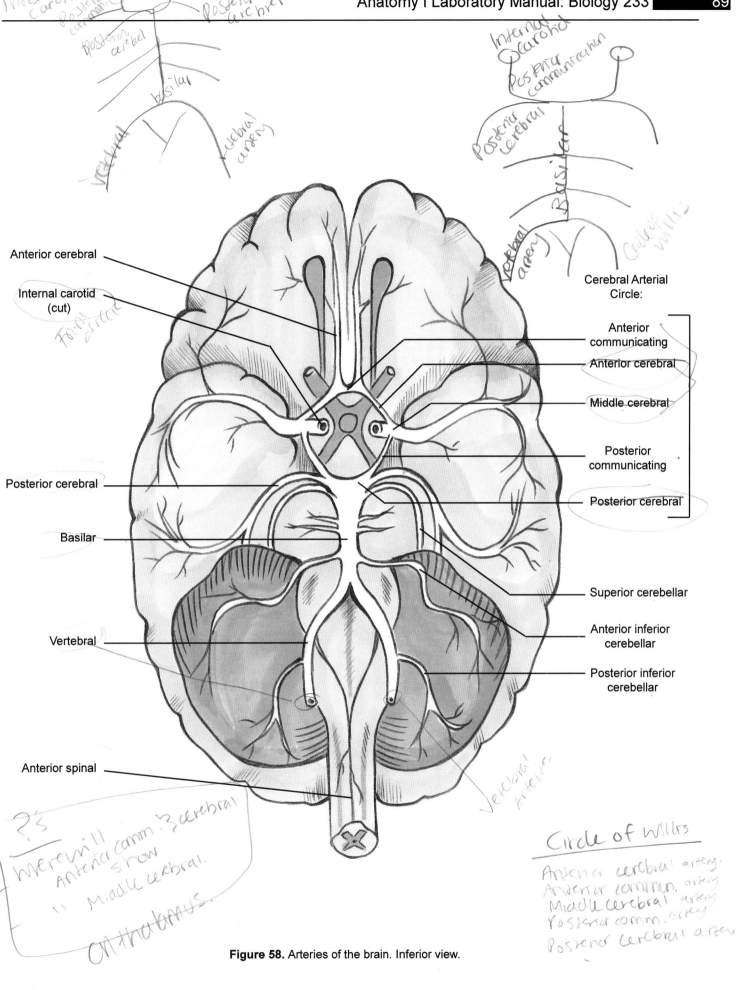

Figure 58. Arteries of the brain. Inferior view.

Anterior cerebral

Internal carotid (cut)

Posterior cerebral

Basilar

Vertebral

Anterior spinal

Cerebral Arterial Circle:

Anterior communicating

Anterior cerebral

Middle cerebral

Posterior communicating

Posterior cerebral

Superior cerebellar

Anterior inferior cerebellar

Posterior inferior cerebellar

Brain:
General Features & Regions

The brain is encased by the skull (bone), held in place by the meninges (connective tissue), and surrounded by cerebrospinal fluid (liquid). The adult brain weighs approximately 1.5 kg (3 – 3.5 lbs) and is made up of 100 billion neurons. The brain requires a constant supply of oxygen and nutrients due to its high metabolic rate – remember, neurotransmission in the brain is constantly taking place. The brain receives nearly 750 ml of blood per minute; flow which aids in oxygen delivery, nutrient supply, and waste removal.

There are **six regions** of the brain: the Cerebrum, Cerebellum, Diencephalon, Mesencephalon, Pons, and Medulla oblongata – the latter three are considered the **brain stem**.

1. **Cerebrum** – largest region, divided into left and right hemispheres by the longitudinal cerebral fissure.

2. **Cerebellum** – second largest part, divided into two hemispheres.

3. **Diencephalon** – located deep to the cerebrum and cerebellum, links the cerebrum with the brain stem. Has several divisions: epithalamus, thalamus, hypothalamus, and pituitary gland.

Brain stem:

4. **Mesencephalon** – also called the midbrain.

5. **Pons** – connects the cerebellum to the brain stem.

6. **Medulla oblongata** – connection point between the brain and the spinal cord.

Cerebrum: Region 1 of 6
Higher mental function (memory storage, thought processes, the ability to reason, the perception of sensory impulses) occurs in the **cerebrum** (SUH-rē-brum; *cerebrum* = brain). The cerebrum makes up approximately 80% of our brain mass and includes **left** and **right hemispheres**, which are partially separated by the **longitudinal cerebral fissure**. The **corpus callosum** is a tract of white matter that unifies the two hemispheres, allowing a sharing of information to take place between the two sides.

There are two layers to the cerebrum; an outer **cerebral cortex** (*cortex* = bark) made up gray matter and an inner **white matter** layer. During fetal development the folds and grooves of the cerebrum develop, they are called **convolutions**. The elevated ridges of the convolutions that create the numerous folds we see when looking at the adult cerebrum are called **cerebral gyri** (singular: JĪ-rus, plural: JĪ-rī; *gyros* = circle), and the shallow depressions are called **cerebral sulci** (singular: SUL-sus or SUL-kus, plural: SUL-sī or SUL- kī; *sulcus* = a furrow or ditch). The infoldings of the convolutions dramatically increase the amount of gray matter surface area in the brain, and thereby increase the amount of nerve cell bodies found here.

Lobes of the Cerebrum

Each of the cerebral hemispheres has four lobes, each with their own primary functions and separated by deep sulci (Figure 59).

Frontal Lobe
The **frontal lobe** makes up the anterior portion of the brain. The **central sulcus** separates it from the parietal lobe, while the **lateral sulcus** separates it from the temporal lobe. The **precentral gyrus** is directly anterior to the central sulcus, and is involved in motor functioning.

Parietal Lobe *[handwritten: motor skins]*
The **parietal lobe** is just posterior to the frontal lobe. The first landmark is the **postcentral gyrus**, located just behind the central sulcus. It is a key sensory area which responds to stimuli from receptors throughout the body. *[handwritten: Sensory]*

Temporal Lobe
The **temporal lobe** is posterior to the frontal lobe and inferior to the parietal lobe. The lateral sulcus separates it from both. The auditory center and memories of auditory and visual experiences are stored here. If the temporal lobe were removed, a subregion called the **insula** (*insula* = island) would be exposed.

Occipital Lobe
The posterior portion of the cerebrum is formed by the **occipital lobe**. It lies superior to the cerebellum, and is separated from it by the tentorium cerebelli. Vision and visual associations are integrated here.

Cerebellum: Region 2 of 6 *[handwritten: Balance + primary motor, posture]*
The **cerebellum** is the brains second largest structure, and is found in the most inferior and posterior portion of the cranium (Figure 59). It consists of left and right hemispheres constricted by a **vermis** (*vermis* = worm). The cerebellum has a thin, outer layer of gray matter called the cerebellar cortex, and a thick, deep layer of white matter. From a sagittal view, the white matter tracts create a branching system called the **arbor vitae** (*arbor* = tree; *vitae* = life). The main function of the cerebellum is to coordinate skeletal muscle contractions. It makes constant impulses to specific motor units to maintain posture, balance, and muscle tone.

Diencephalon: Region 3 of 6
The diencephalon is one of the most important autonomic regions of the brain, which includes the epithalamus, thalamus, hypothalamus, and pituitary gland (Figure 60).

Epithalamus
The dorsal portion of the diencephalon is the location of the **epithalamus**. Here, over the third ventricle, we find a thin lining of the vascular **choroid plexus** where cerebrospinal fluid is made. The small **pineal** (*pinea* = pine cone) **gland** extends from the back and participates in neuroendocrine functioning.

Thalamus
This organ actually consists of a **right** and **left thalamus**, each side within its respective cerebral hemisphere (Figure 61). Its size accounts for the majority of the diencephalon. The thalamus (*thalamus* = inner room) is essentially a relay center for nearly all sensory impulses (excluding smell) to the cerebral cortex.

Hypothalamus
The **hypothalamus** is located below the thalamus, hence its name (*hypo* = below). Although it is small in size, the hypothalamus plays an enormous role in vital autonomic functions – including: endocrine function, cardiovascular regulation, temperature regulation, water and electrolyte balance, emotional responses, and hormone secretion.

Pituitary Gland
The **hypophysis** or **pituitary gland** (*pituita* = phlegm; originally thought to secrete mucus into the nasal cavity… oh those crazy 17th century anatomists, such simple minded people) is attached to the underside of the hypothalamus by a small stalk called the **infundibulum** (*infundibulum* = funnel). It is a marble shaped structure, housed in the sella turcica of the sphenoid bone, which participates in endocrine function. It can be divided into an anterior adenohypophysis, and a posterior neurohypophysis.

Brain Stem

Autonomic functions of the body are coordinated in the brain stem and the nuclei located there (remember: the cell bodies within the CNS usually gather into areas called nuclei, do not confuse this with the nucleus of a cell) before reaching the cerebrum. This attachment point of the brain and spinal cord is where the mesencephalon, pons, and medulla oblongata are located.

Mesencephalon: Region 4 of 6
The **mesencephalon** seems like a long, drawn out word but it is just a fancy way of saying **midbrain** (*meso* = middle; *enkephalos* = in the head). It is located between the diencephalon and the pons (Figure 60). The **mesencephalic aqueduct** (or **cerebral aqueduct**, **aqueduct of Sylvius**, **aqueduct of the midbrain**) connects the 3rd and 4th ventricles of the brain and is located here. The **corpora quadrigemina**, for visual and auditory reflexes, is also located in the midbrain.

Pons: Region 5 of 6
Located anteriorly between the midbrain and the medulla oblongata is an oval shaped bulge called the **pons** (Figure 60). It functions to link the cerebellum with the cerebrum, and the medulla oblongata with the midbrain.

Medulla oblongata: Region 6 of 6
At the level of the foramen magnum (remember that big hole at the bottom of our skull?), the brain stem exits as the **medulla oblongata** (Figure 60). It is a structure that looks like the spinal cord except for the pyramids on the inferior side and an oval shaped structure called the olive on each lateral surface. Within the medulla oblongata is an open space called the **fourth ventricle**, which is continuous with the central canal of the spinal cord. Autonomic centers controlling cardiac functions, blood pressure, and respiration are located in the medulla oblongata. Also, cranial nerves VIII, IX, X, XI, and XII arise from this structure.

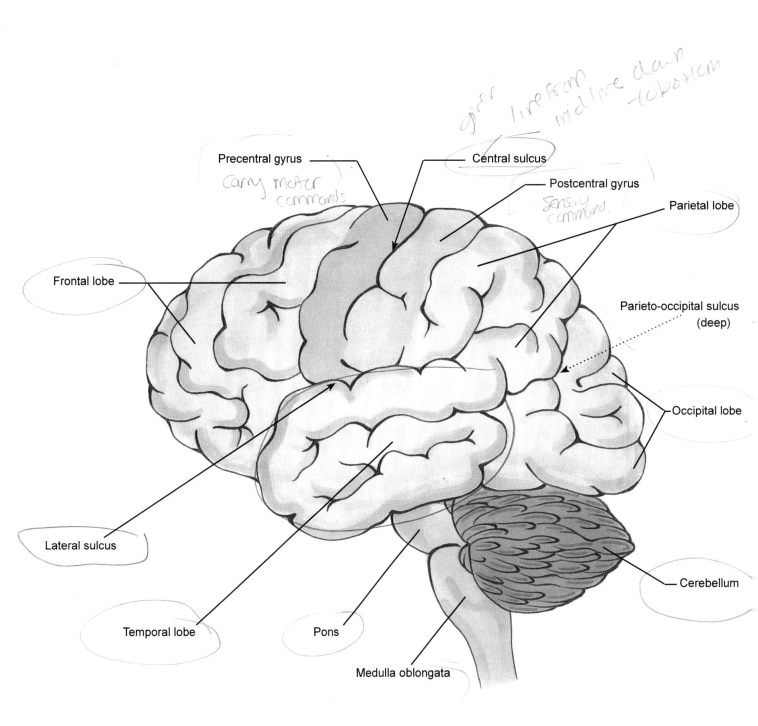

over the from midline clown rotostion

Precentral gyrus

Carry motor commands

Central sulcus

Postcentral gyrus

Sensory command

Parietal lobe

Frontal lobe

Parieto-occipital sulcus (deep)

Occipital lobe

Lateral sulcus

Cerebellum

Temporal lobe

Pons

Medulla oblongata

Figure 59. Lateral view of the brain.

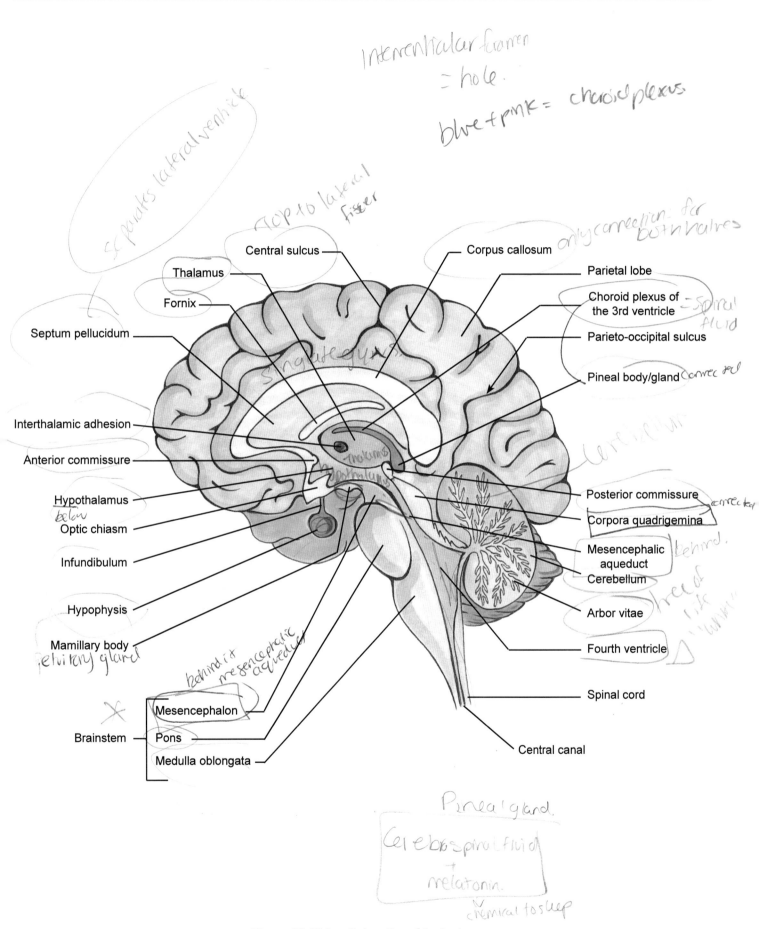

Figure 60. Midsagittal section of the brain.

Central sulcus

Thalamus

Fornix

Septum pellucidum

Interthalamic adhesion

Anterior commissure

Hypothalamus

Optic chiasm

Infundibulum

Hypophysis

Mamillary body

Brainstem

Mesencephalon

Pons

Medulla oblongata

Corpus callosum

Parietal lobe

Choroid plexus of the 3rd ventricle

Parieto-occipital sulcus

Pineal body/gland

Posterior commissure

Corpora quadrigemina

Mesencephalic aqueduct

Cerebellum

Arbor vitae

Fourth ventricle

Spinal cord

Central canal

(handwritten annotations) Interventricular foramen = hole. blue + pink = choroid plexus. separates lateral ventricle. top to lateral fissure. only connection for both halves. = spinal fluid. cingulate gyrus. Thalamus. hypothalamus. below. pituitary gland. behind it mesencephalic aqueduct. connected. cerebellum. behind. tree of life. Pineal gland. Cerebrospinal fluid + melatonin. chemical to sleep.

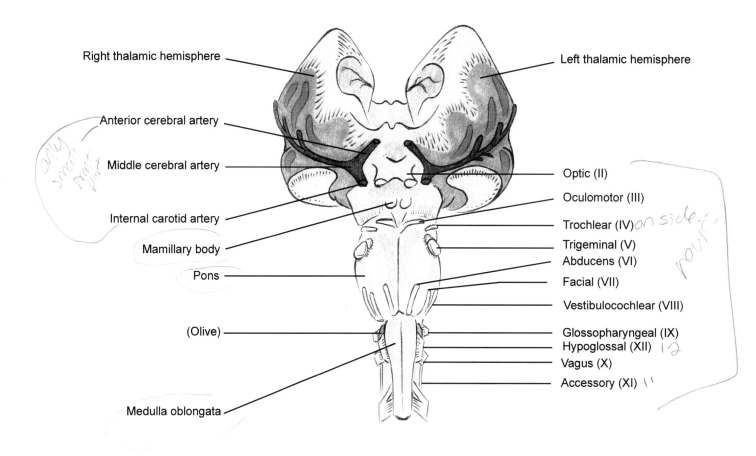

Right thalamic hemisphere
Left thalamic hemisphere
Anterior cerebral artery
Middle cerebral artery
Internal carotid artery
Mamillary body
Pons
(Olive)
Medulla oblongata

only small part

Optic (II)
Oculomotor (III)
Trochlear (IV) on side-?
Trigeminal (V)
Abducens (VI)
Facial (VII)
Vestibulocochlear (VIII)
Glossopharyngeal (IX)
Hypoglossal (XII) 12
Vagus (X)
Accessory (XI) 11

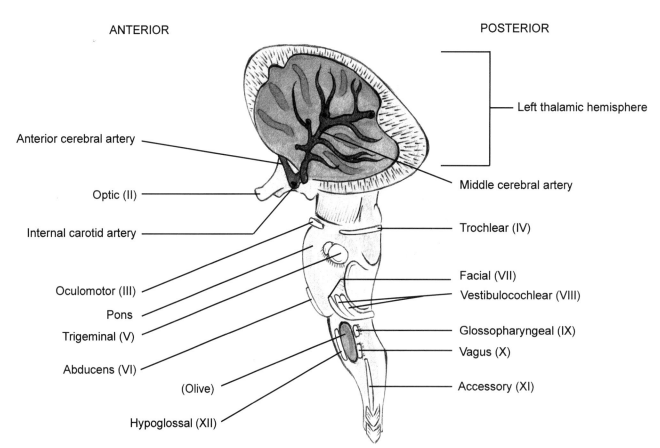

ANTERIOR
POSTERIOR

Anterior cerebral artery
Left thalamic hemisphere
Optic (II)
Internal carotid artery
Middle cerebral artery
Trochlear (IV)
Oculomotor (III)
Facial (VII)
Pons
Vestibulocochlear (VIII)
Trigeminal (V)
Glossopharyngeal (IX)
Abducens (VI)
Vagus (X)
(Olive)
Accessory (XI)
Hypoglossal (XII)

Figure 61. Diencephalon and brain stem with cranial nerves. Anterior and lateral view.

Ventricular System of the Brain

The **ventricles** of the brain are continuous with one another, as well as the central canal of the spinal cord (Figure 62). These hollow chambers are filled with cerebrospinal fluid and lined with a type of neuroglia called ependymal cells. The paired **lateral ventricles** are large C-shaped chambers found deep within each cerebral hemisphere. Anteriorly, they are separated only by a thin membrane, the **septum pellucidum** (*septum* = wall; *pellucidus* = allowing passage of light). From there, the **interventricular foramen** (or **foramen of Monro**; John Monro, American 1858-1910) leads to the **third ventricle** that lies within the diencephalon. The third ventricle then continues down via the **mesencephalic aqueduct** (or **cerebral aqueduct, aqueduct of the midbrain, aqueduct of Sylvius**; Jacobus Sylvius, French 1478-1555) which is a canal-like passageway leading to the **fourth ventricle**. The fourth ventricle is then continuous with the central canal of the spinal cord.

Cranial Nerves

Twelve (12) **cranial nerves** (Table 20) lead straight from the brain and brainstem to sensory structures and muscles of the head and neck (Figure 61 and 63). Some of these cranial nerves also distribute to organs in the thoracic cavity and upper portions of the gastrointestinal tract.

Cranial nerve I is the **olfactory** (*olfacere* = smell out) nerve, which is involved in the sense of smell. Cranial nerve II is the **optic** (*optica* = see) nerve, which sends visual information from the retina of the eye to the brain. Cranial nerves III, IV, and VI, the **oculomotor, trochlear** (*trochos* = a pully, wheel), and **abducens** nerve, respectively, control muscles that move the eyeball and eyelid. Cranial nerve V, the **trigeminal** (*trigeminus* = three born together) nerve, has three subdivisions: the ophthalmic, maxillary, and mandibular branches, which transmit impulses for chewing and sensory sensations from the head.

Cranial nerve VII, the **facial** nerve, provides motor fibers for facial movements and sensory sensations for the taste buds. Cranial nerve VIII, the **vestibulocochlear** (*vestibulum* = chamber; *cochlea* = snail shell) carries impulses for hearing and helps to maintain balance and equilibrium. Cranial nerve IX is the **glossopharyngeal** (*glossa* = tongue; *pharynx* = throat) nerve, a mixed nerve that innervates part of the tongue and pharynx which stimulates the swallowing reflex and the secretion of saliva. Cranial nerve X, the **vagus** (*vagus* = wandering) nerve, is the longest of the cranial nerves whose branches innervate muscles of the pharynx, larynx, respiratory tract, lungs, heart, esophagus, and abdominal viscera, with the exception of the lower portion of the large intestine. Cranial nerve XI, the (spinal) **accessory** nerve innervates skeletal muscles of the soft palate, pharynx, larynx, which contract while swallowing; as well as the sternocleidomastoid and trapezius muscles that move the head, neck, and shoulders. Cranial nerve XII, the **hypoglossal** (*hypo* = under; *glossa* = tongue) nerve, accounts for the coordinated movements of our tongue – food manipulation, swallowing, and speech.

Table 20. Cranial Nerves

NUMBER	NAME	NUMBER	NAME
I	Olfactory	VII	Facial
II	Optic	VIII	Vestibulocochlear
III	Oculomotor	IX	Glossopharyngeal
IV	Trochlear	X	Vagus
V	Trigeminal	XI	Accessory
VI	Abducens	XII	Hypoglossal

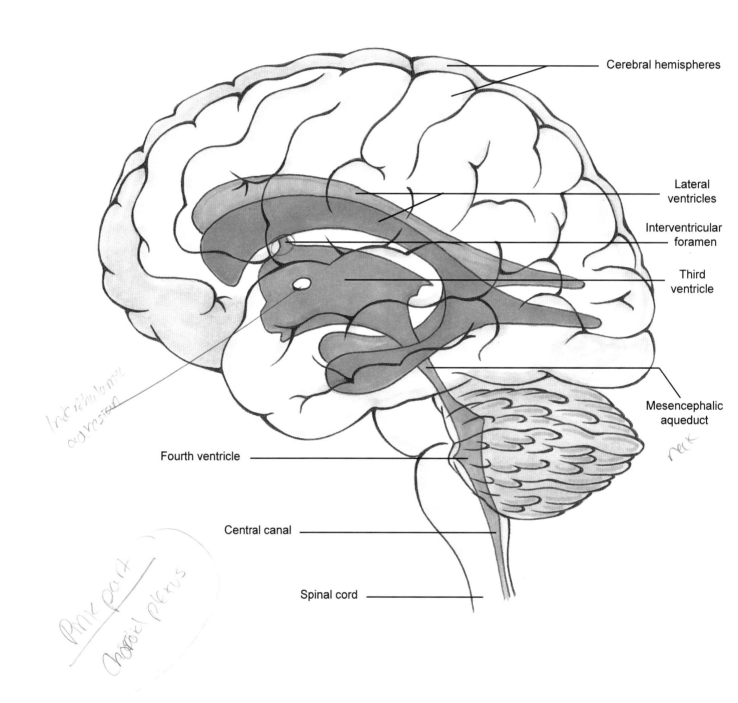

Cerebral hemispheres

Lateral ventricles

Interventricular foramen

Third ventricle

Mesencephalic aqueduct

Fourth ventricle

Central canal

Spinal cord

Figure 62. Ventricles of the brain. Lateral view.

Oh oh on to touch And
Feel very Good velvet
And heaven

long   short

3-32
4-33

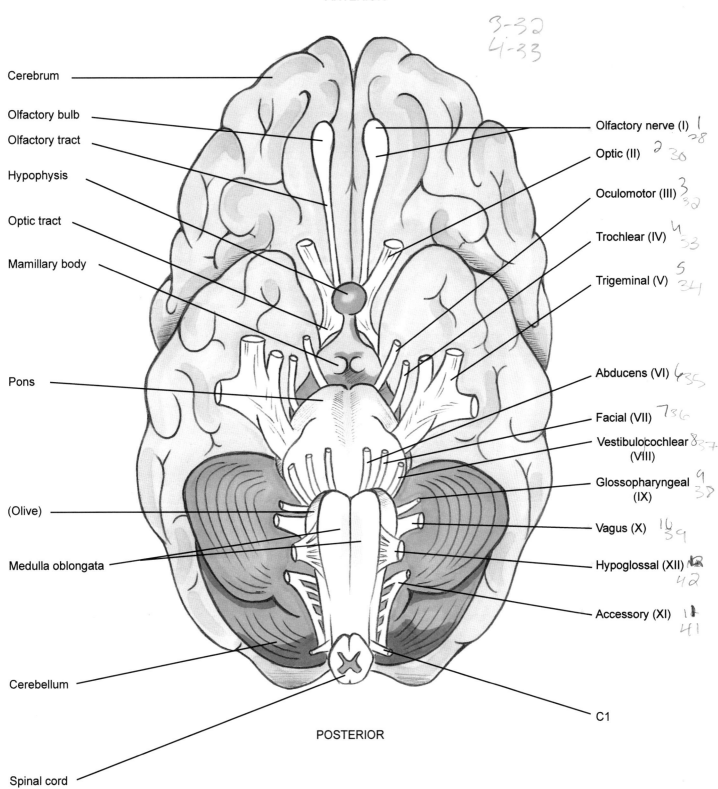

ANTERIOR

Cerebrum

Olfactory bulb

Olfactory tract

Hypophysis

Optic tract

Mamillary body

Pons

(Olive)

Medulla oblongata

Cerebellum

Spinal cord

Olfactory nerve (I) 1 28

Optic (II) 2 30

Oculomotor (III) 3 32

Trochlear (IV) 4 33

Trigeminal (V) 5 34

Abducens (VI) 6 35

Facial (VII) 7 36

Vestibulocochlear 8 37
(VIII)

Glossopharyngeal 9 38
(IX)

Vagus (X) 10 39

Hypoglossal (XII) 12 42

Accessory (XI) 11 41

C1

POSTERIOR

Figure 63. Origins of the cranial nerves. Inferior view.

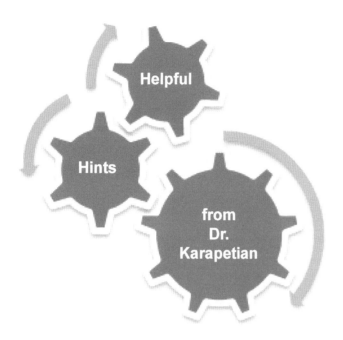

To remember the twelve cranial nerves in order, try memorizing the following mnemonic: On Occasion Our Trusty Truck Acts Funny, Very Good Vehicle Any How …yes, I know the ending is grammatically incorrect; "Anyhow" is actually one word, but do you want my help with this or not?

Spinal Cord

he spinal cord provides a two-way conduction pathway to and from the brain. It is encased within the vertebral column and extends from the foramen magnum down to the region of the first (L1) or second (L2) lumbar vertebra. The adult spinal cord is about 45 cm (18 in) in length and reaches a maximum of about 1.8 cm (3/4 of an inch) thick. Similar to the brain, the spinal cord is protected by bone, meninges, and cerebrospinal fluid. On the outer region there is a large **epidural space** filled with fat and blood vessels (Figure 56).

etween the arachnoid and pia mater is an area called the **subarachnoid space** where cerebrospinal fluid circulates. If a clinical sample of cerebrospinal fluid is necessary, a lumbar puncture or tap below the region of L1 is typically where it is drawn from because the spinal cord is no longer apparent, just the nerve roots that can shift away from the inserted needle.

As you look at its lower portion you will see the spinal cord ends at a tapered, cone-shaped structure called the **conus medullaris** (*konos* = cone). From there, an extension of the pia mater called the **filum terminale** (*filum* = thread; *terminale* = final) continues downward to attach on the posterior surface of the coccyx (Figure 64 and 66). This attachment anchors the spinal cord in place at the bottom, whereas along its length, the spinal cord is anchored to the walls of the vertebral canal by the **denticulate ligaments**.

e have **31 pairs of spinal nerves** which exit the vertebral column via the **intervertebral foramina** and travel to specific regions of the body (Figure 65). The 31 pairs of spinal nerves are grouped as: 8 cervical, 12 thoracic, 5 lumbar, 5 sacral, and 1 coccygeal. The first pair of cervical nerves emerges between the occipital bone of the skull and C1, the atlas vertebra. The second through the seventh pairs of cervical nerves emerge above the vertebrae for which they are named, however the eighth pair of cervical nerves pass between the seventh cervical and first thoracic vertebrae. The remaining pairs of spinal nerves emerge below the vertebrae for which they are named.

The spinal cord has both **cervical** and **lumbar enlargements** where nerves that serve the upper and lower limbs originate (Figure 64). Remember, the spinal cord does not reach the end of the vertebral column, so the lumbar and sacral nerve roots have a way to travel before they reach their intervertebral foramina. This assembly of nerve roots at the end of the spinal cord is termed the **cauda equina** (*cauda* = tail; *equus* = horse) because it looks like a horse's tail.

Spinal Cord: Cross-Sectional Anatomy

 transverse section of the spinal cord reveals its oval shaped structures with a relatively deep **anterior median fissure**, and a shallow **posterior median sulcus** (Figure 55). These grooves run the length of the spinal cord and partially distinguish its left and right symmetry. Gray matter is located inside, while white matter makes up the outside of the cord.

The butterfly shape of gray matter crosses at the **gray commissure** (*commissura* = union) and encloses the **central canal**. There are two **posterior (dorsal) horns** and two **anterior (ventral) horns**; and in the thoracic and lumbar regions there are also gray **lateral horns** present. Interesting note: The amount of ventral gray matter present indicates the amount of skeletal muscle innervated at that particular level of the spinal cord. This is why we see the biggest anterior horns at the limb-innervating cervical and lumbar regions of the spinal cord, and also why the spinal cord has enlargements in those regions.

Sensory information from the body (periphery) enters the spinal cord via the **dorsal roots**, where nerve cell bodies are found in the enlarged region of the dorsal root called the **dorsal root ganglion**. Motor neurons of the anterior gray horns, conversely, send axons out the **ventral roots** to skeletal muscles. The dorsal and ventral roots unite laterally to the spinal cord and form **spinal nerves**, which are considered mixed nerves because of the sensory (in) and motor (out) information they carry.

Nerve Plexuses

There are four plexuses of spinal nerves: the cervical, brachial, lumbar, and sacral (Figure 65 and 66). Each one is described below.

1. **Cervical Plexus**
   The **cervical plexus** can be found deep in the neck region; it is formed by the anterior rami (*ramus* = branch) of the first four cervical nerves (**C1 – C4**) and a **part of C5**. The cervical plexus has branches that innervate the skin, neck muscles, and areas of the head and shoulders. An important component of this plexus comes from the third, fourth, and fifth cervical nerves which unite to form the **phrenic** (FREN- ik) **nerve**, which innervates the diaphragm and is necessary for breathing.

2. **Brachial Plexus**
   The **brachial plexus** is formed by the anterior rami (branches) of the spinal nerves **C5 – T1**. The brachial plexus extends down, passing over the first rib, behind the clavicle, and eventually gives rise to the nerves in the upper extremities (the arms).

3. **Lumbar Plexus**
   The **lumbar plexus** is formed by the fibers of **T12**, and the interior rami (branches) of the **L1 – L4** spinal nerves. Nerves that arise here innervate the lower abdomen and the anterior and medial portions of the lower extremities (the thighs and legs). The **femoral nerve** is a large branch that arises here and divides into smaller branches to give motor impulses and sensations to the hip and thigh.

4. **Sacral Plexus**
   The **sacral plexus** is caudal to the lumbar plexus, and is formed by the anterior rami (branches) of spinal nerves **L4, L5**, and **S1 – S4**. Nerves that arise here innervate the lower back, pelvis, posterior surface of the thigh and leg, and the dorsal and plantar surfaces of the foot. An important component here is the **sciatic** (sī-AT-ik; *sciaticus* = hip joint) **nerve**, the largest nerve in the body. It travels from the pelvis, through the greater sciatic notch of the os coxae, and extends downward on the posterior side of the thigh. Interestingly, two nerves (the tibial and common fibular) combine to compose the sciatic nerve. The next time your leg "falls asleep" you can thank your sciatic nerve… a real bundle of joy that feels like a thousand pins poking your leg. This tingling feeling occurs because you compressed the nerve, probably from sitting too long on a hard surface – ouch!

Thoracolumbar Regional Structures: Sympathetic Division

 reganglionic fibers that exit the spinal cord by way of the ventral root enter a connecting paravertebral (chain) ganglion that forms part of the **sympathetic trunk** or **chain**. This chain appears like a string of pearls that borders each side of the vertebral column.

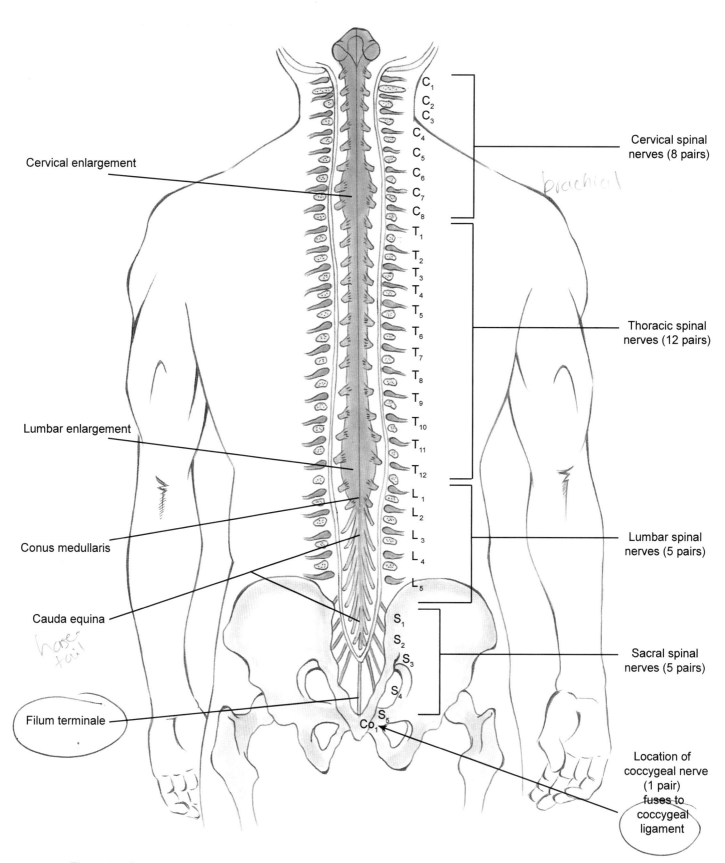

Cervical enlargement

Lumbar enlargement

Conus medullaris

Cauda equina

Filum terminale

C_1
C_2
C_3
C_4
C_5
C_6
C_7
C_8
T_1
T_2
T_3
T_4
T_5
T_6
T_7
T_8
T_9
T_{10}
T_{11}
T_{12}
L_1
L_2
L_3
L_4
L_5
S_1
S_2
S_3
S_4
S_5
Co_1

brachial

Cervical spinal
nerves (8 pairs)

Thoracic spinal
nerves (12 pairs)

Lumbar spinal
nerves (5 pairs)

Sacral spinal
nerves (5 pairs)

Location of
coccygeal nerve
(1 pair)
fuses to
coccygeal
ligament

horse tail

Figure 64. Spinal cord, posterior view. Vertebral arches of bones removed to display spinal cord and its nerve roots.

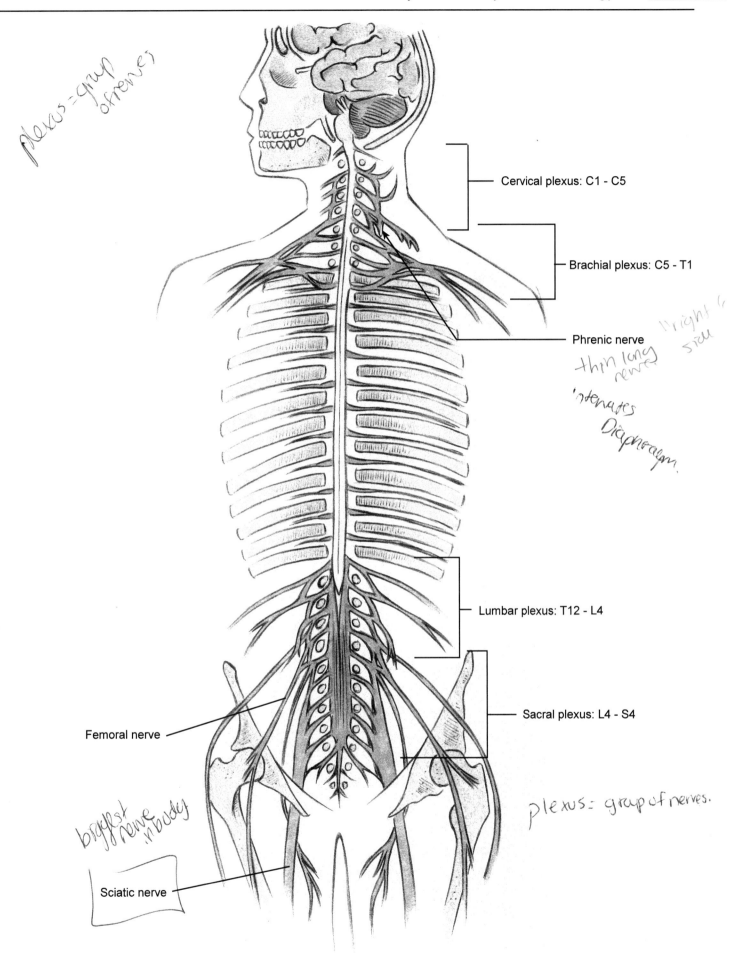

plexus = group of nerves

Cervical plexus: C1 - C5

Brachial plexus: C5 - T1

Phrenic nerve

"right"
side
thin long
nerve
innervates
Diaphragm.

Lumbar plexus: T12 - L4

Femoral nerve

Sacral plexus: L4 - S4

biggest nerve
in body

Sciatic nerve

plexus = group of nerves.

Figure 65. Nerve plexuses, anterior view.

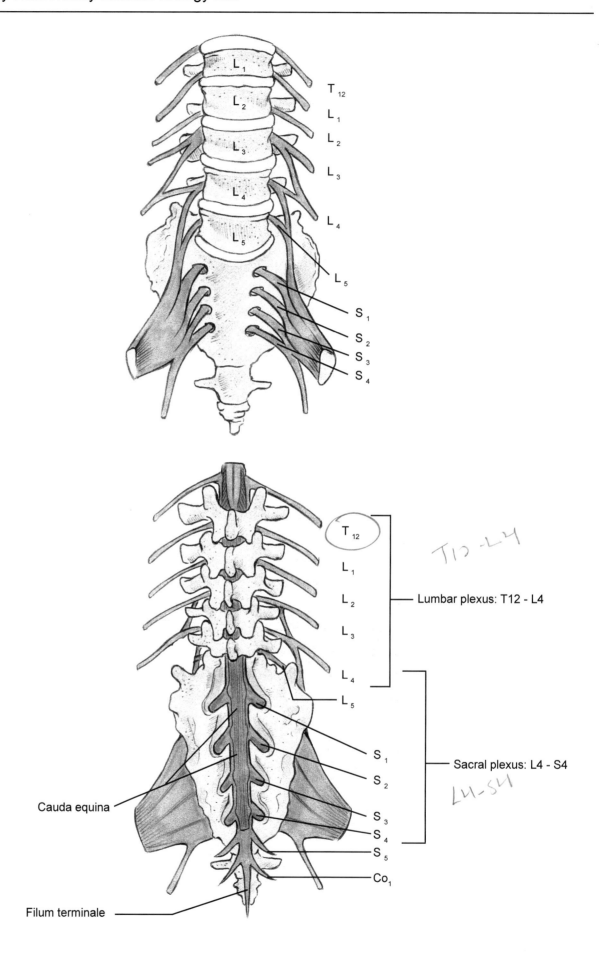

Figure 66. Lumbar and sacral plexuses. Anterior and posterior view.

REFERENCE TABLES

Table 21. Root words indicating position

ENGLISH	GREEK	LATIN
Around, surrounding	peri-	circum-
Above	epi-	supra-
Below	hypo-	infra-
Inside	endo-	intra-
Left	levo-	laevo- , sinistr-
Middle	mes(o)-	medi-
Right	dexi(o)-	dextr(o)-

Table 22 . Root words indicating quantity

ENGLISH	GREEK	LATIN
Double	diplo-	dupli-
Equal	iso-	equi-
Few	oligo-	pauci-
Half	hemi-	semi-
Many	poly-	multi-
Twice	dis-	bis-

Element	Definition	Example	Element	Definition	Example
a-, an-	absence of	atrophy	cyst-	bladder, sac	cystitis
ab-	away from	abduct	cyt(o)-	cell	cytokine
abdomino-	relating to the abdomen	abdominal	de-	away from, down	descend
-ac, -acal	pertaining to	cardiac	dent-	teeth	dentist
acousio-	relating to hearing	acoustic	derm-	skin	dermatology
acro-	extremity, topmost	acromion process	di-	two	diarthrotic
ad-	toward, to	adduct	dis-	separation, apart	dissection
adeno-	relating to a gland	adinocarcinoma	dys-	bad, difficult	dysentery
adipo-	relating to fat	adipocyte	ec-	out, away	eccrine
-al	pertaining to	abdominal	ecto-	outer, outside	ectopic pregnancy
alb-	white or pale color	linea alba	-ectomy	surgical removal	mastectomy
-algia	pain	neuralgia	-emia	blood condition	anemia
ambi-	both of two	ambidextrous	encephalo-	brain	encephalogram
andro-	pertaining to a man	android	endo-	inside, within	endoderm
angio-	blood vessel	angiology	entero-	intestine	gastroenterology
ante-	in front of another thing	antebrachium	epi-	upon, after, in addition	epicardium
anti-	opposed to	antibody	episio-	pubic region	episiotomy
apo-	separated from	apoptosis	erythro-	red	erythrocyte
arterio-	pertaining to an artery	arteriole	eu-	true, good, well, new	eukaryote
arthr-	joints	arthritis	ex-	out of, away from	excise
-ase	enzyme	lactase	exo-	outside	exoskeleton
-asthenia	weakness	myasthenia gravis	extra-	beyond, in addition	extracellular
auto-	self	autoimmune	fasci-	band	fascia
bi-	two	bipedal	-ferent	carry	efferent
bio-	life	biology	for-	opening	foramen
blast(o)-	generative or germ bud	blastocyst	-form	having the form of	cuneiform
brachi(o)-	arm	brachialis	gastro-	pertains to stomach	gastric bypass
brady-	slow	bradycardia	-gen	born in, from	pathogen
bronch(i)-	bronchus	bronchitis	-genic	formative, producing	carcinogenic
bucc(o)-	cheek	buccal cavity	gloss-	tongue	glossopharyngeal
capill-	hair	blood capillaries	glyco-	sugar	glycolysis
caput-	head	decapitate	-gnosis	knowledge	diagnosis
carcin-	cancer	carcinoma	-gram	record	angiogram
cardi(o)-	heart	cardiology	-graph	recording instrument	electrocardiograph
cata-	down	catabolism	gyn-	female sex	gynecomastia
cele-	abdominal	celiac artery	halluc-	to wander in mind	hallucination
cephal-	head	hydrocephalus	hemat-	blood	hematology
cerebro-	brain	cerebrospinal	hemi-	half	cerebral hemisphere
cervic-	neck	cervix of uterus	hepat-	liver	hepatic portal
chemo-	chemistry, drug	chemotherapy	hetero-	other, different	heterogeneous
chiasm-	crossing	optic chiasm	histo-	web, tissue	histology
chole-	bile	cholecystokinin	homo-	same, alike	homologous
chondr-	cartilage	chondrogenic	hydro-	water	hydrophobic
chrom-	color	chromosomes stain darkly	hyper-	extreme, above normal	hypertension
cili-	small hair	ciliated epithelium	hypo-	under, below normal	hypoglycemia
circum-	around	circumcision	hystero-	womb, uterus	hysterectomy
co-	together	concentric	-ia	condition	hypoglycemia
contra	against	contraindicate	-iatry	medical specialty	psychiatry
cornu-	horn	stratum corneum	-ic	pertaining to	hepatic artery
cost-	rib	intercostal	idio-	self, distinct	idiopathic
crani-	skull	craniology	ilio-	ilium	iliosacral
-crine	to secrete	endocrine	infra-	below	infraspinatus
cutic-	skin	subcutaneous	inter-	between, among	interosseous
cyan-	blue	cyanosis, blue skin	intra-	inside, within	intracellular

Element	Definition	Example	Element	Definition	Example
iso-	equal	isotonic	-plegia	paralysis	paraplegia
-itis	inflammation	tonsillitis	-pnea	to breathe	apnea
-ium	structure, tissue	pericardium	pneumato-	breathing, lungs	pneumonia
isch-	restriction	ischemia	pod-	foot	podiatry
karyo-	nucleus	eukaryote	-poiesis	formation of	hemopoiesis
kerato-	cornea (eye or skin)	keratoscope	poly-	many, much	polydactyly
kinesio-	movement	kinesiology	post-	after, behind	postpartum
labi-	lip	labiodental	pre-	before in place or time	prenatal
lacri-	tear	lacrimal gland	prim-	first	primary
laryng-	larynx, voice box	larynx	pro-	before in place or time	prosect
later-	side	lateral	proct-	anus	proctology
leuk-	white	leukocyte	pseudo-	false	pseudostratified
lip-	fat	liposuction	psycho-	mental	psychology
-logy	science of	cardiology	-ptosis	falling	apoptosis
-lysis	dissolve, separation	hemolysis	quad-	four	quadriceps
macro-	large, great	macrophage	re-	again, back	repolarization
mal-	bad, abnormal	malignant	rect-	straight	rectus femoris
medi-	middle	medial	reno-	kidney	renal
mega-	large	megakaryocyte	retro-	backward, behind	retrovirus
meso-	middle	mesoderm	rhin-	nose	rhinitis
meta-	after, behind	metacarpal	-rrhage	excessive flow	hemorrhage
micro-	small	microscope	-rrhea	flow, discharge	diarrhea
mono-	single	monocyte	sanguin-	blood	sanguiferous
morph-	form, shape	morphology	sarc-	fleshlike	sarcoplasm
myo-	muscle	myocardium	scolio-	twisted	scoliosis
narc-	numb, sleep	narcolepsy, narcotic	-scope	instrument for viewing	stethoscope
necro-	death	necrosis	semi-	one-half	semilunar
neo-	new	neoplasm	somato-	pertaining to the body	somatosensory
nephro-	kidney	nephrology	steno-	narrow	stenography
neuro-	nerve	neurolemma	-stomy	surgical opening	tracheostomy
oc-	against	occlusion	sub-	beneath	subcutaneous
-oid	resemblance to	sigmoid	super-	in excess, above	superior vena cava
oligo-	little, few	oligodendrocyte	supra-	above, over	suprarenal
-oma	tumor	lymphoma	sym- (syn-)	similarity, together	synapse
oo-	egg	oogenesis	tachy-	fast	tachycardia
ophthalmo-	eye	ophthalmology	tele-	far	telencephalon
or-	mouth	oral	tens-	stretch	tensor fascia latae
ortho-	straight, correct	orthodontist	therm-	heat	homeotherm
-ory	pertaining to	sensory	thorac-	chest	thoracic cavity
osteo-	bone	osteoporosis	thrombo-	clot, lump	thrombocyte
oto-	ear	otolith	-tomy	cut, incision	splenectomy
ovo-	egg	ovum	tox-	poison	toxic
para-	alongside of	paranasal	trans-	across, over	transversus abdominis
path(o)-	disease	pathology	-trophy	development	hypertrophy
-pathy	abnormality, disease	sociopathy	-tropic	turning toward	gonadotropic
ped-	children	pediatrician	ultra-	beyond, excessive	ultrasonic
-penia	deficiency	osteopenia	uni-	one	unicellular
per-	through	percutaneous	uro-	urine, urinary system	urology
peri-	near, around	pericardium	utero-	womb	uterus
phag-	eat, devour	phagocyte	vaso-	blood vessel	vasoconstriction
pharmaco-	drug, medication	pharmacology	ventro-	belly, stomach cavities	ventrodorsal
-phil	attraction for	neutrophil	viscer-	internal organs	viscera
phlebo-	vein	phlebotomy	xeno-	foreign, different	xenograft
-phobe	exaggerated fear	hydrophobic	zoo-	animal	zoology
-plasty	surgical repair	rhinoplasty	zygo-	union	zygote

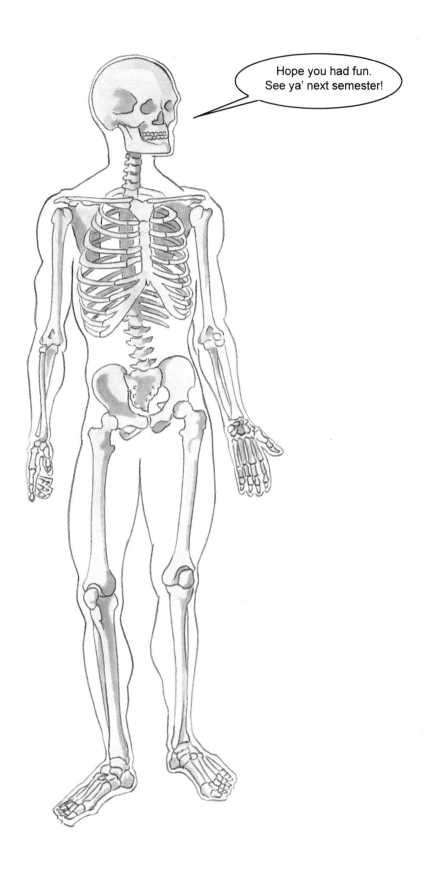